SpringerBriefs in Applied Sciences and Technology

SpringerBriefs present concise summaries of cutting-edge research and practical applications across a wide spectrum of fields. Featuring compact volumes of 50–125 pages, the series covers a range of content from professional to academic.

Typical publications can be:

- A timely report of state-of-the art methods
- An introduction to or a manual for the application of mathematical or computer techniques
- A bridge between new research results, as published in journal articles
- A snapshot of a hot or emerging topic
- An in-depth case study
- A presentation of core concepts that students must understand in order to make independent contributions

SpringerBriefs are characterized by fast, global electronic dissemination, standard publishing contracts, standardized manuscript preparation and formatting guidelines, and expedited production schedules.

On the one hand, **SpringerBriefs in Applied Sciences and Technology** are devoted to the publication of fundamentals and applications within the different classical engineering disciplines as well as in interdisciplinary fields that recently emerged between these areas. On the other hand, as the boundary separating fundamental research and applied technology is more and more dissolving, this series is particularly open to trans-disciplinary topics between fundamental science and engineering.

Indexed by EI-Compendex, SCOPUS and Springerlink.

More information about this series at http://www.springer.com/series/8884

Ricardo Almeida · Dina Tavares
Delfim F. M. Torres

The Variable-Order
Fractional Calculus
of Variations

 Springer

Ricardo Almeida
Department of Mathematics
University of Aveiro
Aveiro, Portugal

Delfim F. M. Torres
Department of Mathematics
University of Aveiro
Aveiro, Portugal

Dina Tavares
Polytechnic Institute of Leiria
Leiria, Portugal

ISSN 2191-530X   ISSN 2191-5318   (electronic)
SpringerBriefs in Applied Sciences and Technology
ISBN 978-3-319-94005-2   ISBN 978-3-319-94006-9   (eBook)
https://doi.org/10.1007/978-3-319-94006-9

Library of Congress Control Number: 2018945050
Mathematics Subject Classification (2010): 49K05, 49K21, 26A33, 34A08

Printed on acid-free paper

This Springer imprint is published by the registered company Springer International Publishing AG part of Springer Nature
The registered company address is: Gewerbestrasse 11, 6330 Cham, Switzerland

Preface

This book intends to deepen the study of the fractional calculus, giving special emphasis to variable-order operators.

Fractional calculus is a recent field of mathematical analysis, and it is a generalization of integer differential calculus, involving derivatives and integrals of real or complex order [17, 27]. The first note about this idea of differentiation, for non-integer numbers, dates back to 1695, with a famous correspondence between Leibniz and L'Hôpital. In a letter, L'Hôpital asked Leibniz about the possibility of the order n in the notation $d^n y/dx^n$, for the nth derivative of the function y, to be a non-integer, $n = 1/2$. Since then, several mathematicians investigated this approach, like Lacroix, Fourier, Liouville, Riemann, Letnikov, Grünwald, Caputo, and contributed to the grown development of this field. Currently, this is one of the most intensively developing areas of mathematical analysis as a result of its numerous applications. The first book devoted to the fractional calculus was published by Oldham and Spanier in 1974, where the authors systematized the main ideas, methods, and applications about this field [18].

In the recent years, fractional calculus has attracted the attention of many mathematicians, but also some researchers in other areas like physics, chemistry, and engineering. As it is well known, several physical phenomena are often better described by fractional derivatives [13, 22, 30]. This is mainly due to the fact that fractional operators take into consideration the evolution of the system, by taking the global correlation, and not only local characteristics. Moreover, integer-order calculus sometimes contradict the experimental results, and therefore, derivatives of fractional order may be more suitable [14].

In 1993, Samko and Ross devoted themselves to investigate operators when the order α is not a constant during the process, but variable on time: $\alpha(t)$ [35]. An interesting recent generalization of the theory of fractional calculus is developed to allow the fractional order of the derivative to be non-constant, depending on time [9, 23, 24]. With this approach of variable-order fractional calculus, the non-local properties are more evident, and numerous applications have been found in physics, mechanics, control, and signal processing [10, 15, 25, 26, 28, 31, 42].

Although there are many definitions of fractional derivative, the most commonly used are the Riemann–Liouville, the Caputo, and the Grünwald–Letnikov derivatives. For more about the development of fractional calculus, we suggest [17, 18, 27, 34, 35].

One difficult issue that usually arises when dealing with such fractional operators is the extreme difficulty in solving analytically such problems [7, 44]. Thus, in most cases, we do not know the exact solution for the problem, and one needs to seek a numerical approximation. Several numerical methods can be found in the literature, typically applying some discretization over time or replacing the fractional operators by a proper decomposition [7, 44].

Recently, new approximation formulas were given for fractional constant-order operators, with the advantage that higher-order derivatives are not required to obtain a good accuracy of the method [6, 30, 31]. These decompositions only depend on integer-order derivatives, and by replacing the fractional operators that appear in the problem by them, one leaves the fractional context ending up in the presence of a standard problem, where numerous tools are available to solve them [2].

The first goal of this book is to extend such decompositions to Caputo fractional problems of variable order. For three types of Caputo derivatives with variable order, we obtain approximation formulas for the fractional operators and respective upper bounds for the errors.

Then, we focus our attention on a special operator introduced by Malinowska and Torres: the combined Caputo fractional derivative, which is an extension of the left and the right fractional Caputo derivatives [19]. Considering $\alpha, \beta \in (0, 1)$ and $\gamma \in [0, 1]$, the combined Caputo fractional derivative operator ${}^C D_\gamma^{\alpha,\beta}$ is a convex combination of the left and the right Caputo fractional derivatives, defined by

$$
{}^C D_\gamma^{\alpha,\beta} = \gamma \, {}^C_a D_t^\alpha + (1 - \gamma) \, {}^C_t D_b^\beta.
$$

We consider this fractional operator with variable fractional order, i.e., the combined Caputo fractional derivative of variable order:

$$
{}^C D_\gamma^{\alpha(\cdot,\cdot),\beta(\cdot,\cdot)} x(t) = \gamma_1 \, {}^C_a D_t^{\alpha(\cdot,\cdot)} x(t) + \gamma_2 \, {}^C_t D_b^{\beta(\cdot,\cdot)} x(t),
$$

where $\gamma = (\gamma_1, \gamma_2) \in [0, 1]^2$, with γ_1 and γ_2 not both zero. With this fractional operator, we study different types of fractional calculus of variations problems, where the Lagrangian depends on the referred derivative.

The calculus of variations is a mathematical subject that appeared formally in the seventeenth century, with the solution to the Brachistochrone problem, that deals with the extremization (minimization or maximization) of functionals [43]. Usually, functionals are given by an integral that involves one or more functions or/and its derivatives. This branch of mathematics has proved to be relevant because of the numerous applications existing in real situations.

The fractional variational calculus is a recent mathematical field that consists in minimizing or maximizing functionals that depend on fractional operators (integrals or/and derivatives). This subject was introduced by Riewe in 1996, where the author generalizes the classical calculus of variations, by using fractional derivatives, and allows to obtain conservations laws with nonconservative forces such as friction [33, 34]. Later appeared several works on various aspects of the fractional calculus of variations and involving different fractional operators, like the Riemann–Liouville, the Caputo, the Grünwald–Letnikov, the Weyl, the Marchaud or the Hadamard fractional derivatives [1, 3–5, 8, 11, 12, 16]. For the state of the art of the fractional calculus of variations, we refer the readers to the books [2, 20, 21].

Specifically, here we study some problems of the calculus of variations with integrands depending on the independent variable t, an arbitrary function x and a fractional derivative $^{C}D_{\gamma}^{\alpha(\cdot,\cdot),\beta(\cdot,\cdot)}x$. The endpoint of the cost integral, as well the terminal state, is considered to be free. The fractional problem of the calculus of variations consists in finding the maximizers or minimizers to the functional

$$\mathcal{J}(x,T) = \int_{a}^{T} L\left(t, x(t), {}^{C}D_{\gamma}^{\alpha(\cdot,\cdot),\beta(\cdot,\cdot)}x(t)\right) dt + \varphi(T, x(T)),$$

where $^{C}D_{\gamma}^{\alpha(\cdot,\cdot),\beta(\cdot,\cdot)}x(t)$ stands for the combined Caputo fractional derivative of variable fractional order, subject to the boundary condition $x(a) = x_a$. For all variational problems presented here, we establish necessary optimality conditions and transversality optimality conditions.

The book is organized in two parts, as follows. In the first part, we review the basic concepts of fractional calculus (Chap. 1) and of the fractional calculus of variations (Chap. 2). In Chap. 1, we start with a brief overview about fractional calculus and an introduction to the theory of some special functions in fractional calculus. Then, we recall several fractional operators (integrals and derivatives) definitions, and some properties of the considered fractional derivatives and integrals are introduced. In the end of this chapter, we review integration by parts formulas for different operators. Chapter 2 presents a short introduction to the classical calculus of variations and review different variational problems, like the isoperimetric problems or problems with variable endpoints. In the end of this chapter, we introduce the theory of the fractional calculus of variations and some fractional variational problems with variable order.

In the second part, we systematize some new recent results on variable-order fractional calculus of [37–41]. In Chap. 3, considering three types of fractional Caputo derivatives of variable order, we present new approximation formulas for those fractional derivatives and prove upper-bound formulas for the errors. In Chap. 4, we introduce the combined Caputo fractional derivative of variable-order and corresponding higher-order operators. Some properties are also given. Then, we prove fractional Euler–Lagrange equations for several types of fractional problems of the calculus of variations, with or without constraints.

Aveiro, Portugal   Ricardo Almeida
Leiria, Portugal   Dina Tavares
Aveiro, Portugal   Delfim F. M. Torres

References

1. Agrawal OP (2002) Formulation of Euler–Lagrange equations for fractional variational problems. J Math Anal Appl 272:368–379
2. Almeida R, Pooseh S, Torres DFM (2015) Computational methods in the fractional calculus of variations. Imperial College Press, London
3. Almeida R (2016) Fractional variational problems depending on indefinite integrals and with delay. Bull Malays Math Sci Soc 39(4):1515–1528
4. Askari H, Ansari A (2016) Fractional calculus of variations with a generalized fractional derivative. Fract Differ Calc 6:57–72
5. Atanacković TM, Konjik S, Pilipović S (2008) Variational problems with fractional derivatives: Euler–Lagrange equations. J Phys A 41(9):095201, 12 pp
6. Atanacković TM, Janev M, Pilipović S, Zorica D (2013) An expansion formula for fractional derivatives of variable order. Cent Eur J Phys 11(10):1350—1360
7. Atangana A, Cloot AH (2013) Stability and convergence of the space fractional variable-order Schrödinger equation. Adv Difference Equ 80(1):10 pp
8. Baleanu D (2008) New applications of fractional variational principles. Rep Math Phys 61(2):199–206
9. Chen S, Liu F, Burrage K (2014) Numerical simulation of a new two–dimensional variable-order fractional percolation equation in non-homogeneous porous media. Comput Math Appl 67(9):1673–1681
10. Coimbra CFM, Soon CM, Kobayashi MH (2005) The variable viscoelasticity operator. Annalen der Physik 14(6):378—389
11. Fraser C (1992) Isoperimetric problems in variational calculus of Euler and Lagrange. Historia Mathematica 19:4–23
12. Georgieva B, Guenther RB (2002) First Noether-type theorem for the generalized variational principle of Herglotz. Topol Methods Nonlinear Anal 20(2):261–273
13. Herrmann R (2013) Folded potentials in cluster physics–a comparison of Yukawa and Coulomb potentials with Riesz fractional integrals. J Phys A 46(40):405203, 12 pp
14. Hilfer R (2000) Applications of fractional calculus in physics. World Sci. Publishing, River Edge, NJ, USA
15. Ingman D, Suzdalnitsky J (2004) Control of damping oscillations by fractional differential operator with time–dependent order. Comput Meth Appl Mech Eng 193(52):5585–5595
16. Jarad F, Abdeljawad T, Baleanu D (2010) Fractional variational principles with delay within Caputo derivatives. Rep Math Phys 65(1):17–28
17. Kilbas AA, Srivastava HM, Trujillo JJ (2006) Theory and applications of fractional differential equations. Elsevier, Amsterdam
18. Mainardi F (2010) Fractional calculus and waves in linear viscoelasticity. Imperial College Press, London

19. Malinowska AB, Torres DFM (2010) Fractional variational calculus in terms of a combined Caputo derivative. In: Podlubny I, Vinagre Jara BM, Chen YQ, Feliu Batlle V, Tejado Balsera I (eds) Proceedings of FDA'10, The 4th IFAC workshop on fractional differentiation and its applications, Badajoz, Spain, October 18–20, 2010 :Article no. FDA10-084, 6 pp
20. Malinowska AB, Odzijewicz T, Torres DFM (2015) Advanced methods in the fractional calculus of variations. Springer Briefs in Applied Sciences and Technology, Springer, Cham
21. Malinowska AB, Torres DFM (2012) Introduction to the fractional calculus of variations. Imperial College Press, London
22. Odzijewicz T, Malinowska AB, Torres DFM (2012) Fractional calculus of variations in terms of a generalized fractional integral with applications to physics. Abstr Appl Anal 2012:Art. ID 871912, 24 pp
23. Odzijewicz T, Malinowska AB, Torres DFM (2012) Variable order fractional variational calculus for double integrals. In: Proceedings of the 51st IEEE Conference on Decision and Control, December 10–13, 2012, Maui, Hawaii: Art. no. 6426489, 6873–6878
24. Odzijewicz T, Malinowska AB, Torres DFM (2013) Fractional variational calculus of variable order. In: Advances in harmonic analysis and operator theory, 291–301, Oper Theory Adv Appl, Birkhäuser/Springer Basel AG, Basel
25. Odzijewicz T, Malinowska AB, Torres DFM (2013) Noether's theorem for fractional variational problems of variable order. Cent Eur J Phys 11(6):691–701
26. Ostalczyk PW, Duch P, Brzeziń ski DW, Sankowski D (2015) Order functions selection in the variable-, fractional-order PID controller. Advances in Modelling and Control of Non–integer–Order Systems, Lecture Notes in Electrical Engineering 320:159–170
27. Podlubny I (1999) Fractional differential equations. Academic Press, San Diego, CA
28. Pooseh S, Almeida R, Torres DFM (2012) Approximation of fractional integrals by means of derivatives. Comput Math Appl 64(10):3090–3100
29. Pooseh S, Almeida R, Torres DFM (2013) Numerical approximations of fractional derivatives with applications. Asian J Control 15(3):698–712
30. Ramirez LES, Coimbra CFM (2011) On the variable order dynamics of the nonlinear wake caused by a sedimenting particle. Phys D 240(13):1111–1118
31. Rapaić MR, Pisano A (2014) Variable-order fractional operators for adaptive order and parameter estimation. IEEE Trans Automat Control 59(3):798–803
32. Riewe F (1996) Nonconservative Lagrangian and Hamiltonian mechanics. Phys Rev E(3) 53 (2):1890–1899
33. Riewe F (1997) Mechanics with fractional derivatives. Phys Rev E(3) 55(3):3581–3592
34. Samko SG, Ross B (1993) Integration and differentiation to a variable fractional order. Integral Transform Spec Funct 1(4):277–300
35. Samko SG, Kilbas AA, Marichev OI (1993) Fractional Integrals and Derivatives. translated from the 1987 Russian original, Gordon and Breach, Yverdon
36. Tavares D, Almeida R, Torres DFM (2015) Optimality conditions for fractional variational problems with dependence on a combined Caputo derivative of variable order. Optimization 64(6):1381–1391
37. Tavares D, Almeida R, Torres DFM (2016) Caputo derivatives of fractional variable order: numerical approximations. Commun Nonlinear Sci Numer Simul 35:69–87
38. Tavares D, Almeida R, Torres DFM (2017) Constrained fractional variational problems of variable order. IEEE/CAA J Automatica Sinica 4(1):80–88
39. Tavares D, Almeida R, Torres DFM (2018) Fractional Herglotz variational problem of variable order. Disc Contin Dyn Syst Ser S 11(1):143–154
40. Tavares D, Almeida R, Torres DFM (2018) Combined fractional variational problems of variable order and some computational aspects. J Comput Appl Math 339:374–388
41. Valério D, Costa JS (2013) Variable order fractional controllers. Asian J Control 15(3):648–657

42. van Brunt B (2004) The calculus of variations. Universitext, Springer, New York
43. Zheng B (2012) (G'/G)-expansion method for solving fractional partial differential equations in the theory of mathematical physics. Commun Theor Phys 58(5):623–630
44. Zhuang P, Liu F, Anh V, Turner I (2009) Numerical methods for the variable-order fractional advection-diffusion equation with a nonlinear source term. SIAM J. Numer. Anal. 47 (3):1760–1781

Acknowledgements

This work was supported by Portuguese funds through the *Center for Research and Development in Mathematics and Applications* (CIDMA), and the *Portuguese Foundation for Science and Technology* (FCT), within project UID/MAT/04106/2013.

Any comments or suggestions related to the material here contained are more than welcome, and may be submitted by post or by electronic mail to the authors:

Ricardo Almeida (e-mail: ricardo.almeida@ua.pt)
Center for Research and Development in Mathematics and Applications
Department of Mathematics, University of Aveiro
3810-193 Aveiro, Portugal

Dina Tavares (e-mail: dtavares@ipleiria.pt)
ESECS, Polytechnic Institute of Leiria
2410–272 Leiria, Portugal

Delfim F. M. Torres (e-mail: delfim@ua.pt)
Center for Research and Development in Mathematics and Applications
Department of Mathematics, University of Aveiro
3810-193 Aveiro, Portugal

Contents

Chapter 1
Fractional Calculus

In this chapter, a brief introduction to the theory of fractional calculus is presented. We start with a historical perspective of the theory, with a strong connection with the development of classical calculus (Sect. 1.1). Then, in Sect. 1.2, we review some definitions and properties about a few special functions that will be needed. We end with a review on fractional integrals and fractional derivatives of noninteger order and with some formulas of integration by parts, involving fractional operators (Sect. 1.3).

The content of this chapter can be found in some classical books on fractional calculus, for example, Almeida et al. [5], Kilbas et al. [12], Malinowska and Torres [21], Podlubny [32], Samko et al. [37].

1.1 Historical Perspective

Fractional Calculus (FC) is considered as a branch of mathematical analysis which deals with the investigation and applications of integrals and derivatives of arbitrary order. Therefore, FC is an extension of the integer-order calculus that considers integrals and derivatives of any real or complex order [12, 37], i.e., unify and generalize the notions of integer-order differentiation and n-fold integration.

FC was born in 1695 with a letter that L'Hôpital wrote to Leibniz, where the derivative of order 1/2 is suggested [29]. After Leibniz had introduced in his publications the notation for the nth derivative of a function y,

$$\frac{d^n y}{dx^n},$$

L'Hôpital wrote a letter to Leibniz to ask him about the possibility of a derivative of integer order to be extended in order to have a meaning when the order is a fraction:

© The Author(s), under exclusive license to Springer International
Publishing AG, part of Springer Nature 2019
R. Almeida et al., *The Variable-Order Fractional Calculus of Variations*, SpringerBriefs
in Applied Sciences and Technology, https://doi.org/10.1007/978-3-319-94006-9_1

"What if n be $1/2$?" [34]. In his answer, dated on September 30, 1695, Leibniz replied that "This is an apparent paradox from which, one day, useful consequences will be drawn" and, today, we know it is truth. Then, Leibniz still wrote about derivatives of general order and in 1730, Euler investigated the result of the derivative when the order n is a fraction. But, only in 1819, with Lacroix, appeared the first definition of fractional derivative based on the expression for the nth derivative of the power function. Considering $y = x^m$, with m a positive integer, Lacroix developed the nth derivative

$$\frac{d^n y}{dx^n} = \frac{m!}{(m-n)!} x^{m-n}, \quad m \geq n,$$

and using the definition of Gamma function, for the generalized factorial, he got

$$\frac{d^n y}{dx^n} = \frac{\Gamma(m+1)}{\Gamma(m-n+1)} x^{m-n}.$$

Lacroix also studied the following example, for $n = \frac{1}{2}$ and $m = 1$:

$$\frac{d^{1/2} x}{dx^{1/2}} = \frac{\Gamma(2)}{\Gamma(3/2)} x^{\frac{1}{2}} = \frac{2\sqrt{x}}{\sqrt{\pi}}. \tag{1.1}$$

Since then, many mathematicians, like Fourier, Abel, Riemann, Liouville, among others, contributed to the development of this subject. One of the first applications of fractional calculus appear in 1823 by Niels Abel, through the solution of an integral equation of the form

$$\int_0^t (t-\tau)^{-\alpha} x(\tau) d\tau = k,$$

used in the formulation of the tautochrone problem [1, 34].

Different forms of fractional operators have been introduced along time, like the Riemann–Liouville, the Grünwald–Letnikov, the Weyl, the Caputo, the Marchaud, or the Hadamard fractional derivatives [12, 29, 30, 32]. The first approach is the Riemann–Liouville, which is based on iterating the classical integral operator n times and then considering the Cauchy's formula where $n!$ is replaced by the Gamma function and hence the fractional integral of noninteger order is defined. Then, using this operator, some of the fractional derivatives mentioned above are defined.

During three centuries, FC was developed but as a pure theoretical subject of mathematics. In recent times, FC had an increasing of importance due to its applications in various fields, not only in mathematics, but also in physics, mechanics, engineering, chemistry, biology, finance, and other areas of science [10, 11, 15, 17, 28, 31, 39, 40]. In some of these applications, many real-world phenomena are better described by noninteger-order derivatives, if we compare with the usual integer-order calculus. In fact, fractional-order derivatives have unique characteristics that may model certain dynamics more efficiently. Firstly, we can consider any real order for the derivatives, and thus we are not restricted to integer-order derivatives only; secondly, they are nonlocal operators, in opposite to the usual derivatives, containing memory. With the memory property, one can take into account the past of the processes. Signal processing, modeling, and control are some areas that have been the object of more intensive publishing in the last decades.

In most applications of the FC, the order of the derivative is assumed to be fixed along the process, that is, when determining what is the order $\alpha > 0$ such that the solution of the fractional differential equation $D^\alpha y(t) = f(t, y(t))$ better approaches the experimental data, we consider the order α to be a fixed constant. Of course, this may not be the best option, since trajectories are a dynamic process, and the order may vary. More interesting possibilities arise when one considers the order α of the fractional integrals and derivatives not constant during the process but to be a function $\alpha(t)$, depending on time. Then, we may seek what is the best function $\alpha(\cdot)$ such that the variable-order fractional differential equation $D^{\alpha(t)} y(t) = f(t, y(t))$ better describes the process under study. This approach is very recent. One such fractional calculus of variable order was introduced in Samko and Ross [36]. Afterward, several mathematicians obtained important results about variable-order fractional calculus, and some applications appeared, like in mechanics, in the modeling of linear and nonlinear viscoelasticity oscillators, and in other phenomena where the order of the derivative varies with time. See, for instance, Almeida and Torres [4], Atanacković and Pilipović [6], Coimbra [8], Odzijewicz et al. [26], Ramirez and Coimbra [33], Samko [35], Sheng et al. [38].

The most common fractional operators considered in the literature take into account the past of the process. They are usually called left fractional operators. But in some cases, we may be also interested in the future of the process, and the computation of $\alpha(t)$ to be influenced by it. In that case, right fractional derivatives are then considered. Recently, in some works, the main goal is to develop a theory where both fractional operators are taken into account. For that, some combined fractional operators are introduced, like the symmetric fractional derivative, the Riesz fractional integral and derivative, the Riesz–Caputo fractional derivative, and the combined Caputo fractional derivative that consists in a linear combination of the left and right fractional operators. For studies with fixed fractional order, see Klimek [13], Malinowska and Torres [20–22].

Due to the growing number of applications of fractional calculus in science and engineering, numerical approaches are being developed to provide tools for solving such problems. At present, there are already vast studies on numerical approximate formulas [14, 16]. For example, for numerical modeling of time-fractional diffusion equations, we refer the reader to Fu et al. [9].

1.2   Special Functions

Before introducing the basic facts on fractional operators, we recall four types of
functions that are important in Fractional Calculus: the *Gamma*, *Psi*, *Beta*, and
Mittag-Leffler functions. Some properties of these functions are also recalled.

Definition 1 The Euler Gamma function is an extension of the factorial function to
real numbers, and it is defined by

$$\Gamma(t) = \int_0^\infty \tau^{t-1} \exp(-\tau) \, d\tau, \quad t > 0.$$

For example, $\Gamma(1) = 1$, $\Gamma(2) = 1$, and $\Gamma(3/2) = \frac{\sqrt{\pi}}{2}$. For positive integers n,
we get $\Gamma(n) = (n-1)!$. We mention that other definitions for the Gamma function
exist, and it is possible to define it for complex numbers, except for the nonpositive
integers.

The Gamma function is considered the most important Eulerian function used
in fractional calculus, because it appears in almost every fractional integral and
derivative definitions. A basic but fundamental property of Γ, that we will use later,
is obtained using integration by parts:

$$\Gamma(t+1) = t \, \Gamma(t).$$

Definition 2 The Psi function is the derivative of the logarithm of the Gamma function:

$$\Psi(t) = \frac{d}{dt} \ln(\Gamma(t)) = \frac{\Gamma'(t)}{\Gamma(t)}.$$

The follow function is used sometimes for convenience to replace a combination
of Gamma functions. It is important in FC because it shares a form that is similar to
the fractional derivative or integral of many functions, particularly power functions.

Definition 3 The Beta function B is defined by

$$B(t, u) = \int_0^1 s^{t-1}(1 - s)^{u-1} ds, \quad t, u > 0.$$

This function satisfies an important property:

$$B(t, u) = \frac{\Gamma(t)\Gamma(u)}{\Gamma(t+u)}.$$

With this property, it is obvious that the Beta function is symmetric, i.e.,

$$B(t, u) = B(u, t).$$

The next function is a direct generalization of the exponential series, and it was defined by the mathematician Mittag-Leffler in 1903 [32].

Definition 4 Let $\alpha > 0$. The function E_α defined by

$$E_\alpha(t) = \sum_{k=0}^{\infty} \frac{t^k}{\Gamma(\alpha k + 1)}, \quad t \in \mathbb{R},$$

is called the one-parameter Mittag-Leffler function.

For $\alpha = 1$, this function coincides with the series expansion of e^t, i.e.,

$$E_1(t) = \sum_{k=0}^{\infty} \frac{t^k}{\Gamma(k + 1)} = \sum_{k=0}^{\infty} \frac{t^k}{k!} = e^t.$$

While linear ordinary differential equations present in general the exponential function as a solution, the Mittag-Leffler function occurs naturally in the solution of fractional-order differential equations [12]. For this reason, in recent times, the Mittag-Leffler function has become an important function in the theory of the fractional calculus and its applications.

It is also common to represent the Mittag-Leffler function in two arguments. This generalization of Mittag-Leffler function was studied by Wiman in 1905 [17].

Definition 5 The two-parameter function of the Mittag-Leffler type with parameters $\alpha, \beta > 0$ is defined by

$$E_{\alpha,\beta}(t) = \sum_{k=0}^{\infty} \frac{t^k}{\Gamma(\alpha k + \beta)}, \quad t \in \mathbb{R}.$$

If $\beta = 1$, this function coincides with the classical Mittag-Leffler function, i.e., $E_{\alpha,1}(t) = E_\alpha(t)$.

1.3 Fractional Integrals and Derivatives

In this section, we recall some definitions of fractional integral and fractional differential operators that include all we use throughout this book. In the end, we present some integration by parts formulas because they have a crucial role in deriving Euler–Lagrange equations.

1.3.1 Classical Operators

As it was seen in Sect. 1.1, there are more than one way to generalize integer-order operations to the noninteger case. Here, we present several definitions and properties about fractional operators, omitting some details about the conditions that ensure the existence of such fractional operators.

In general, the fractional derivatives are defined using fractional integrals. We present only two fractional integral operators, but there are several known forms of the fractional integrals.

Let $x : [a, b] \to \mathbb{R}$ be an integrable function and $\alpha > 0$ a real number.

Starting with Cauchy's formula for a n-fold iterated integral, given by

$$
\begin{aligned}
{}_a I_t^n x(t) &= \int_a^t d\tau_1 \int_a^{\tau_1} d\tau_2 \cdots \int_a^{\tau_{n-1}} x(\tau_n) d\tau_n \\
&= \frac{1}{(n-1)!} \int_a^t (t-\tau)^{n-1} x(\tau) d\tau,
\end{aligned}
\tag{1.2}
$$

where $n \in \mathbb{N}$, Liouville and Riemann defined fractional integration, generalizing equation (1.2) to noninteger values of n and using the definition of Gamma function Γ. With this, we introduce two important concepts: the left and the right Riemann–Liouville fractional integrals.

Definition 6 We define the left and right Riemann–Liouville fractional integrals of order α, respectively, by

$$
{}_a I_t^\alpha x(t) = \frac{1}{\Gamma(\alpha)} \int_a^t (t-\tau)^{\alpha-1} x(\tau) d\tau, \quad t > a
$$

and

$$
{}_t I_b^\alpha x(t) = \frac{1}{\Gamma(\alpha)} \int_t^b (\tau-t)^{\alpha-1} x(\tau) d\tau, \quad t < b.
$$

The constants a and b determine, respectively, the lower and upper boundary of the integral domain. Additionally, if x is a continuous function, as $\alpha \to 0$, ${}_a I_t^\alpha = {}_t I_b^\alpha = I$, with I the identity operator, i.e., ${}_a I_t^\alpha x(t) = {}_t I_b^\alpha x(t) = x(t)$.

We present the second fractional integral operator, introduced by J. Hadamard in 1892 [12].

Definition 7 We define the left and right Hadamard fractional integrals of order α, respectively, by

$$
{}_a J_t^\alpha x(t) = \frac{1}{\Gamma(\alpha)} \int_a^t \left(\ln \frac{t}{\tau} \right)^{\alpha-1} \frac{x(\tau)}{\tau} d\tau, \quad t > a
$$

and

$$_t J_b^\alpha x(t) = \frac{1}{\Gamma(\alpha)} \int_t^b \left(\ln \frac{\tau}{t} \right)^{\alpha-1} \frac{x(\tau)}{\tau} d\tau, \quad t < b.$$

The three most frequently used definitions for fractional derivatives are the Grünwald-Letnikov, the Riemann–Liouville, and the Caputo fractional derivatives [29, 32]. Other definitions were introduced by other mathematicians, as for instance Weyl, Fourier, Cauchy, and Abel.

Let $x \in AC([a, b]; \mathbb{R})$ be an absolutely continuous function on the interval $[a, b]$, and α a positive real number. Using Definition 6 of Riemann–Liouville fractional integrals, we define the left and the right Riemann–Liouville and Caputo derivatives as follows.

Definition 8 We define the left and right Riemann–Liouville fractional derivatives of order $\alpha > 0$, respectively, by

$$_a D_t^\alpha x(t) = \left(\frac{d}{dt} \right)^n {}_a I_t^{n-\alpha} x(t)$$

$$= \frac{1}{\Gamma(n-\alpha)} \left(\frac{d}{dt} \right)^n \int_a^t (t-\tau)^{n-\alpha-1} x(\tau) d\tau, \quad t > a$$

and

$$_t D_b^\alpha x(t) = \left(-\frac{d}{dt} \right)^n {}_t I_b^{n-\alpha} x(t)$$

$$= \frac{(-1)^n}{\Gamma(n-\alpha)} \left(\frac{d}{dt} \right)^n \int_t^b (\tau-t)^{n-\alpha-1} x(\tau) d\tau, \quad t < b,$$

where $n = [\alpha] + 1$.

The following definition was introduced in Caputo [7]. The Caputo fractional derivatives, in general, are more applicable and interesting in fields like physics and engineering, for its properties like the initial conditions.

Definition 9 We define the left and right Caputo fractional derivatives of order α, respectively, by

$$_a^C D_t^\alpha x(t) = {}_a I_t^{n-\alpha} \left(\frac{d}{dt} \right)^n x(t)$$

$$= \frac{1}{\Gamma(n-\alpha)} \int_a^t (t-\tau)^{n-\alpha-1} x^{(n)}(\tau) d\tau, \quad t > a$$

and

$$_t^C D_b^\alpha x(t) = {}_t I_b^{n-\alpha} \left(-\frac{d}{dt} \right)^n x(t)$$

$$= \frac{(-1)^n}{\Gamma(n-\alpha)} \int_t^b (\tau-t)^{n-\alpha-1} x^{(n)}(\tau) d\tau, \quad t < b,$$

where $n = [\alpha] + 1$ if $\alpha \notin \mathbb{N}$ and $n = \alpha$ if $\alpha \in \mathbb{N}$.

Obviously, the above-defined operators are linear. From these definitions, it is clear that the Caputo fractional derivative of a constant C is zero, which is false when we consider the Riemann–Liouville fractional derivative. If $x(t) = C$, with C a constant, then we get

$$_a^C D_t^\alpha x(t) = {_t^C} D_b^\alpha x(t) = 0$$

and

$$_a D_t^\alpha x(t) = \frac{C (t - a)^{-\alpha}}{\Gamma(1 - \alpha)}, \quad _t D_b^\alpha x(t) = \frac{C (b - t)^{-\alpha}}{\Gamma(1 - \alpha)}.$$

For this reason, in some applications, Caputo fractional derivatives seem to be more natural than the Riemann–Liouville fractional derivatives.

Remark 1 If α goes to n^-, with $n \in \mathbb{N}$, then the fractional operators introduced above coincide with the standard derivatives:

$$_a D_t^\alpha = {_a^C} D_t^\alpha = \left(\frac{d}{dt} \right)^n$$

and

$$_t D_b^\alpha = {_t^C} D_b^\alpha = \left(-\frac{d}{dt} \right)^n.$$

The Riemann–Liouville fractional integral and differential operators of order $\alpha > 0$ of power functions return power functions, as we can see below.

Lemma 2 *Let x be the power function $x(t) = (t - a)^\gamma$. Then, we have*

$$_a I_t^\alpha x(t) = \frac{\Gamma(\gamma + 1)}{\Gamma(\gamma + \alpha + 1)} (t - a)^{\gamma + \alpha}, \quad \gamma > -1$$

and

$$_a D_t^\alpha x(t) = \frac{\Gamma(\gamma + 1)}{\Gamma(\gamma - \alpha + 1)} (t - a)^{\gamma - \alpha}, \quad \gamma > -1.$$

In particular, if we consider $\gamma = 1$, $a = 0$, and $\alpha = 1/2$, then the left Riemann–Liouville fractional derivative of $x(t) = t$ is $\frac{2\sqrt{t}}{\sqrt{\pi}}$, the same result (1.1) as Lacroix obtained in 1819.

Grünwald and Letnikov, respectively in 1867 and 1868, returned to the original sources and started the formulation by the fundamental definition of a derivative, as a limit,

$$x^{(1)}(t) = \lim_{h \to 0} \frac{x(t + h) - x(t)}{h}$$

and considering the iteration at the nth-order derivative formula:

$$x^{(n)}(t) = \lim_{h \to 0} \frac{1}{h^n} \sum_{k=0}^{n} (-1)^k \binom{n}{k} x(t - kh), \quad n \in \mathbb{N}, \tag{1.3}$$

where $\binom{n}{k} = \frac{n(n-1)(n-2)...(n-k+1)}{k!}$, with $n, k \in \mathbb{N}$, the usual notation for the binomial coefficients. The Grünwald–Letnikov definition of fractional derivative consists in a generalization of (1.3) to derivatives of arbitrary order α [32].

Definition 10 The αth-order Grünwald–Letnikov fractional derivative of function x is given by

$$_a^{GL}D_t^\alpha x(t) = \lim_{h \to 0} h^{-\alpha} \sum_{k=0}^{n} (-1)^k \binom{\alpha}{k} x(t - kh),$$

where $\binom{\alpha}{k} = \frac{\Gamma(\alpha+1)}{k!\Gamma(\alpha-k+1)}$ and $nh = t - a$.

Lemma 3 Let x be the power function $x(t) = (t - a)^\gamma$, where γ is a real number. Then, for $0 < \alpha < 1$ and $\gamma > 0$, we have

$$_a^{GL}D_t^\alpha x(t) = \frac{\Gamma(\gamma + 1)}{\Gamma(\gamma - \alpha + 1)} (t - a)^{\gamma-\alpha}.$$

Definition 11 We define the left and right Hadamard fractional derivatives of order α, respectively, by

$$_a D_t^\alpha x(t) = \left(t \frac{d}{dt} \right)^n \frac{1}{\Gamma(n - \alpha)} \int_a^t \left(\ln \frac{t}{\tau} \right)^{n-\alpha-1} \frac{x(\tau)}{\tau} d\tau$$

and

$$_t D_b^\alpha x(t) = \left(-t \frac{d}{dt} \right)^n \frac{1}{\Gamma(n - \alpha)} \int_t^b \left(\ln \frac{\tau}{t} \right)^{n-\alpha-1} \frac{x(\tau)}{\tau} d\tau,$$

for all $t \in (a, b)$, where $n = [\alpha] + 1$.

Observe that, for all types of derivative operators, if variable t is the time variable, the left fractional derivative of x is interpreted as a past state of the process, while the right fractional derivative of x is interpreted as a future state of the process.

1.3.2 Some Properties of the Caputo Derivative

For $\alpha > 0$ and $x \in AC([a, b]; \mathbb{R})$, the Riemann–Liouville and Caputo derivatives are related by the following formulas [12]:

$$\substack{C\\a}D_t^\alpha x(t) = {}_aD_t^\alpha \left[x(t) - \sum_{k=0}^{n-1} \frac{x^{(k)}(a)(t-a)^k}{k!} \right] \tag{1.4}$$

and

$$\substack{C\\t}D_b^\alpha x(t) = {}_tD_b^\alpha \left[x(t) - \sum_{k=0}^{n-1} \frac{x^{(k)}(b)(b-t)^k}{k!} \right], \tag{1.5}$$

where $n = [\alpha] + 1$ if $\alpha \notin \mathbb{N}$ and $n = \alpha$ if $\alpha \in \mathbb{N}$.

In particular, when $\alpha \in (0, 1)$, the relations (1.4) and (1.5) take the form:

$$\begin{aligned} \substack{C\\a}D_t^\alpha x(t) &= {}_aD_t^\alpha(x(t) - x(a)) \\ &= {}_aD_t^\alpha x(t) - \tfrac{x(a)}{\Gamma(1-\alpha)}(t-a)^{-\alpha} \end{aligned} \tag{1.6}$$

and

$$\begin{aligned} \substack{C\\t}D_b^\alpha x(t) &= {}_tD_b^\alpha(x(t) - x(b)) \\ &= {}_tD_b^\alpha x(t) - \tfrac{x(b)}{\Gamma(1-\alpha)}(b-t)^{-\alpha}. \end{aligned} \tag{1.7}$$

It follows from (1.6) and (1.7) that the left Riemann–Liouville derivative equals the left Caputo fractional derivative in the case $x(a) = 0$ and the analogue holds for the right derivatives under the assumption $x(b) = 0$.

In Theorem 4, we see that the Caputo fractional derivatives provide a left inverse operator to the Riemann–Liouville fractional integration (cf. Lemma 2.21 in Kilbas et al. [12]).

Theorem 4 *Let $\alpha > 0$ and let $x \in C([a, b]; \mathbb{R}^n)$. For Caputo fractional operators, the next rules hold:*

$$\substack{C\\a}D_t^\alpha \, {}_aI_t^\alpha x(t) = x(t)$$

and

$$\substack{C\\t}D_b^\alpha \, {}_tI_b^\alpha x(t) = x(t).$$

The next statement characterizes the composition of the Riemann–Liouville fractional integration operators with the Caputo fractional differentiation operators (cf. Lemma 2.22 in Kilbas et al. [12]).

Theorem 5 *Let $\alpha > 0$. If $x \in AC^n([a, b]; \mathbb{R})$, then*

$$\substack{a}I_t^\alpha \, \substack{C\\a}D_t^\alpha x(t) = x(t) - \sum_{k=0}^{n-1} \frac{x^{(k)}(a)}{k!}(t-a)^k$$

and

$$\substack{t}I_b^\alpha \, \substack{C\\t}D_b^\alpha x(t) = x(t) - \sum_{k=0}^{n-1} \frac{(-1)^k x^{(k)}(b)}{k!}(b-t)^k,$$

with $n = [\alpha] + 1$ if $\alpha \notin \mathbb{N}$ and $n = \alpha$ if $\alpha \in \mathbb{N}$.

In particular, when $\alpha \in (0, 1)$, then

$$_aI_t^\alpha \, {}^C_aD_t^\alpha x(t) = x(t) - x(a) \quad \text{and} \quad _tI_b^\alpha \, {}^C_tD_b^\alpha x(t) = x(t) - x(b).$$

Similarly to the Riemann–Liouville fractional derivative, the Caputo fractional derivative of a power function yields a power function of the same form.

Lemma 6 Let $\alpha > 0$. Then, the following relations hold:

$$_a^C D_t^\alpha (t - a)^\gamma = \frac{\Gamma(\gamma + 1)}{\Gamma(\gamma - \alpha + 1)} (t - a)^{\gamma - \alpha}, \quad \gamma > n - 1$$

and

$$_t^C D_b^\alpha (b - t)^\gamma = \frac{\Gamma(\gamma + 1)}{\Gamma(\gamma - \alpha + 1)} (b - t)^{\gamma - \alpha}, \quad \gamma > n - 1,$$

with $n = [\alpha] + 1$ if $\alpha \notin \mathbb{N}$ and $n = \alpha$ if $\alpha \in \mathbb{N}$.

1.3.3 Combined Caputo Derivative

In this section, we introduce a special operator for our work, the combined Caputo fractional derivative. We extend the notion of the Caputo fractional derivative to the fractional derivative $^C D_\gamma^{\alpha, \beta}$, that involves the left and the right Caputo fractional derivatives, i.e., it combines the past and the future of the process into one single operator.

This operator was introduced in Malinowska and Torres [18], motivated by the idea of the symmetric fractional derivative introduced in Klimek [13]:

Definition 12 Considering the left and right Riemann–Liouville derivatives, the symmetric fractional derivative is given by

$$_a D_b^\alpha x(t) = \frac{1}{2} \left[_a D_t^\alpha x(t) + _t D_b^\alpha x(t) \right]. \tag{1.8}$$

Other combined operators were studied. For example, we have the Riesz and the Riesz–Caputo operators [21].

Definition 13 Let $x : [a, b] \to \mathbb{R}$ be a function of class C^1 and $\alpha \in (0, 1)$. For $t \in [a, b]$, the Riesz fractional integral of order α is defined by

$$_a^R I_b^\alpha x(t) = \frac{1}{2\Gamma(\alpha)} \int_a^b |t - \tau|^{\alpha - 1} x(\tau) d\tau$$

$$= \frac{1}{2} \left[_a I_t^\alpha x(t) + _t I_b^\alpha x(t) \right],$$

and the Riesz fractional derivative of order α is defined by

$$
{}_a^R D_b^\alpha x(t) = \frac{1}{\Gamma(1-\alpha)} \frac{d}{dt} \int_a^b |t - \tau|^{-\alpha} x(\tau) d\tau
$$
$$
= \frac{1}{2} \left[{}_a D_t^\alpha x(t) - {}_t D_b^\alpha x(t) \right].
$$

Definition 14 Considering the left and right Caputo derivatives, Riesz–Caputo fractional derivative is given by

$$
{}_a^{RC} D_b^\alpha x(t) = \frac{1}{\Gamma(1-\alpha)} \int_a^b |t - \tau|^{-\alpha} \frac{d}{d\tau} x(\tau) d\tau
$$
$$
= \frac{1}{2} \left[{}_a^C D_t^\alpha x(t) - {}_t^C D_b^\alpha x(t) \right].
$$

(1.9)

Remark 7 (*Malinowska and Torres* [21]) If α goes to 1, then the fractional derivatives introduced above coincide with the standard derivative:

$$
{}_a^R D_b^\alpha = {}_a^{RC} D_b^\alpha = \frac{d}{dt}.
$$

Similarly to the last operator (1.9), the combined Caputo derivative is a convex combination of the left and the right Caputo fractional derivatives. But, in this operator, we also consider other coefficients for the convex combination besides $1/2$. Moreover, the orders α and β of the left-and right-sided fractional derivatives can be different. Therefore, the combined Caputo derivative is a convex combination of the left Caputo fractional derivative of order α and the right Caputo fractional derivative of order β.

Definition 15 Let $\alpha, \beta \in (0, 1)$ and $\gamma \in [0, 1]$. The combined Caputo fractional derivative operator ${}^C D_\gamma^{\alpha,\beta}$ is defined by

$$
{}^C D_\gamma^{\alpha,\beta} = \gamma \, {}_a^C D_t^\alpha + (1 - \gamma) \, {}_t^C D_b^\beta,
$$

(1.10)

which acts on $x \in AC([a, b]; \mathbb{R})$ in the following way:

$$
{}^C D_\gamma^{\alpha,\beta} x(t) = \gamma \, {}_a^C D_t^\alpha x(t) + (1 - \gamma) \, {}_t^C D_b^\beta x(t).
$$

The operator (1.10) is obviously linear. Observe that

$$
{}^C D_0^{\alpha,\beta} = {}_t^C D_b^\beta \quad \text{and} \quad {}^C D_1^{\alpha,\beta} = {}_a^C D_t^\alpha.
$$

The symmetric fractional derivative and the Riesz fractional derivative are useful tools to describe some nonconservative models. But those types of differentiation do not seem suitable for all kinds of variational problems because they are based on the Riemann–Liouville fractional derivatives and therefore the possibility that admissible trajectories x have continuous fractional derivatives implies that $x(a) = x(b) = 0$ [36]. For more details about the combined Caputo fractional derivative, see Malinowska and Torres [20, 22, 23], Odzijewicz et al. [25].

1.3.4 Variable-Order Operators

Very useful physical applications have given birth to the variable-order fractional calculus, for example, in modeling mechanical behaviors [9, 42]. Nowadays, variable-order fractional calculus is particularly recognized as a useful and promising approach in the modeling of diffusion processes, in order to characterize time-dependent or concentration-dependent anomalous diffusion, or diffusion processes in inhomogeneous porous media [41].

Now, we present the fundamental notions of the fractional calculus of variable order [19]. We consider the fractional order of the derivative and of the integral to be a continuous function of two variables, $\alpha(\cdot, \cdot)$ with domain $[a, b]^2$, taking values on the open interval $(0, 1)$. Let $x : [a, b] \to \mathbb{R}$ be a function.

First, we recall the generalization of fractional integrals for a variable-order $\alpha(\cdot, \cdot)$.

Definition 16 The left and right Riemann–Liouville fractional integrals of order $\alpha(\cdot, \cdot)$ are defined by

$$_aI_t^{\alpha(\cdot,\cdot)}x(t) = \int_a^t \frac{1}{\Gamma(\alpha(t, \tau))}(t - \tau)^{\alpha(t,\tau)-1}x(\tau)d\tau, \quad t > a$$

and

$$_tI_b^{\alpha(\cdot,\cdot)}x(t) = \int_t^b \frac{1}{\Gamma(\alpha(\tau, t))}(\tau - t)^{\alpha(\tau,t)-1}x(\tau)d\tau, \quad t < b,$$

respectively.

We remark that, in contrast to the fixed fractional-order case, variable-order fractional integrals are not the inverse operation of the variable-order fractional derivatives.

For fractional derivatives, we consider two types: the Riemann–Liouville and the Caputo fractional derivatives.

Definition 17 The left and right Riemann–Liouville fractional derivatives of order $\alpha(\cdot, \cdot)$ are defined by

$$\begin{aligned}
_aD_t^{\alpha(\cdot,\cdot)}x(t) &= \frac{d}{dt}{_aI_t^{\alpha(\cdot,\cdot)}}x(t) \\
&= \frac{d}{dt}\int_a^t \frac{1}{\Gamma(1 - \alpha(t, \tau))}(t - \tau)^{-\alpha(t,\tau)}x(\tau)d\tau, \quad t > a
\end{aligned} \tag{1.11}$$

and

$$
{}_tD_b^{\alpha(\cdot,\cdot)}x(t) = -\frac{d}{dt}{}_tI_b^{\alpha(\cdot,\cdot)}x(t)
$$
$$
= \frac{d}{dt}\int_t^b \frac{-1}{\Gamma(1-\alpha(\tau,t))}(\tau-t)^{-\alpha(\tau,t)}x(\tau)d\tau, \quad t < b \tag{1.12}
$$

respectively.

Lemma 8 gives a Riemann–Liouville variable-order fractional integral and fractional derivative for the power function $x(t) = (t-a)^\gamma$, where we assume that the fractional order depends only on the first variable: $\alpha(t,\tau) := \overline{\alpha}(t)$, where $\overline{\alpha} : [a,b] \to (0,1)$ is a given function.

Lemma 8 *Let x be the power function $x(t) = (t-a)^\gamma$. Then, for $\gamma > -1$, we have*

$$
{}_aI_t^{\overline{\alpha}(t)}x(t) = \frac{\Gamma(\gamma+1)}{\Gamma(\gamma+\overline{\alpha}(t)+1)}(t-a)^{\gamma+\overline{\alpha}(t)}
$$

and

$$
{}_aD_t^{\overline{\alpha}(t)}x(t) = \frac{\Gamma(\gamma+1)}{\Gamma(\gamma-\overline{\alpha}(t)+1)}(t-a)^{\gamma-\overline{\alpha}(t)}
$$
$$
-\overline{\alpha}^{(1)}(t)\frac{\Gamma(\gamma+1)}{\Gamma(\gamma-\overline{\alpha}(t)+2)}(t-a)^{\gamma-\overline{\alpha}(t)+1}
$$
$$
\times [\ln(t-a) - \Psi(\gamma-\overline{\alpha}(t)+2) + \Psi(1-\overline{\alpha}(t))].
$$

Definition 18 The left and right Caputo fractional derivatives of order $\alpha(\cdot,\cdot)$ are defined by

$$
{}_a^CD_t^{\alpha(\cdot,\cdot)}x(t) = \int_a^t \frac{1}{\Gamma(1-\alpha(t,\tau))}(t-\tau)^{-\alpha(t,\tau)}x^{(1)}(\tau)d\tau, \quad t > a \tag{1.13}
$$

and

$$
{}_t^CD_b^{\alpha(\cdot,\cdot)}x(t) = \int_t^b \frac{-1}{\Gamma(1-\alpha(\tau,t))}(\tau-t)^{-\alpha(\tau,t)}x^{(1)}(\tau)d\tau, \quad t < b, \tag{1.14}
$$

respectively.

Of course, the fractional derivatives just defined are linear operators.

1.3.5 Generalized Fractional Operators

In this section, we present three definitions of one-dimensional generalized fractional operators that depend on a general kernel, studied by Odzijewicz, Malinowska, and Torres (see, e.g., Malinowska et al. [19], Odzijewicz et al. [24, 28], although Agrawal [2] had introduced these generalized fractional operators).

Let $\Delta := \{(t, \tau) \in \mathbb{R}^2 : a \leq \tau < t \leq b\}$.

Definition 19 Let k^α be a function defined almost everywhere on Δ with values in $\mathbb{R}$. For all $x : [a, b] \rightarrow \mathbb{R}$, the generalized fractional integral operator K_P is defined by

$$K_P[x](t) = \lambda \int_a^t k^\alpha(t, \tau) x(\tau) d\tau + \mu \int_t^b k^\alpha(\tau, t) x(\tau) d\tau, \qquad (1.15)$$

with $P = \langle a, t, b, \lambda, \mu \rangle$, where λ and μ are real numbers.

In particular, if we choose special cases for the kernel, we can obtain standard fractional operators or variable-order.

Remark 9 For special chosen kernels k^α and parameters P, the operator K_P can be reduced to the classical or variable-order Riemann–Liouville fractional integrals:

- Let $k^\alpha(t, \tau) = \frac{1}{\Gamma(\alpha)}(t - \tau)^{\alpha-1}$ and $0 < \alpha < 1$. If $P = \langle a, t, b, 1, 0 \rangle$, then

$$K_P[x](t) = \frac{1}{\Gamma(\alpha)} \int_a^t (t - \tau)^{\alpha-1} x(\tau) d\tau =: {_aI_t^\alpha}[x](t)$$

is the left Riemann–Liouville fractional integral of order α;
if $P = \langle a, t, b, 0, 1 \rangle$, then

$$K_P[x](t) = \frac{1}{\Gamma(\alpha)} \int_t^b (\tau - t)^{\alpha-1} x(\tau) d\tau =: {_tI_b^\alpha}[x](t)$$

is the right Riemann–Liouville fractional integral of order α.
- If $k^\alpha(t, \tau) = \frac{1}{\Gamma(\alpha(t,\tau))}(t - \tau)^{\alpha(t,\tau)-1}$ and $P = \langle a, t, b, 1, 0 \rangle$, then

$$K_P[x](t) = \int_a^t \frac{1}{\Gamma(\alpha(t, \tau))}(t - \tau)^{\alpha(t,\tau)-1} x(\tau) d\tau =: {_aI_t^{\alpha(\cdot,\cdot)}}[x](t)$$

is the left Riemann–Liouville fractional integral of variable-order $\alpha(\cdot, \cdot)$;
if $P = \langle a, t, b, 0, 1 \rangle$, then

$$K_P[x](t) = \int_t^b \frac{1}{\Gamma(\alpha(t, \tau))}(\tau - t)^{\alpha(t,\tau)-1} x(\tau) d\tau =: {_tI_b^{\alpha(\cdot,\cdot)}}[x](t)$$

is the right Riemann–Liouville fractional integral of variable-order $\alpha(\cdot, \cdot)$.

Some other fractional operators can be obtained with the generalized fractional integrals, for example, Hadamard, Riesz, Katugampola fractional operators [2, 19].
The following two new operators, the generalized fractional Riemann–Liouville and Caputo derivatives, are defined as a composition of classical derivatives and generalized fractional integrals.

Definition 20 The generalized fractional derivative of Riemann–Liouville type, denoted by A_P, is defined by

$$A_P = \frac{d}{dt} \circ K_P.$$

Definition 21 The generalized fractional derivative of Caputo type, denoted by B_P, is defined by

$$B_P = K_P \circ \frac{d}{dt}.$$

Considering $k^\alpha(t, \tau) = \frac{1}{\Gamma(1-\alpha)}(t - \tau)^{-\alpha}$ with $0 < \alpha < 1$ and appropriate sets P, these two general kernel operators A_P and B_P can be reduced to the standard Riemann–Liouville and Caputo fractional derivatives, respectively [19].

1.3.6 Integration by Parts

In this section, we summarize formulas of integration by parts because they are important results to find necessary optimality conditions when dealing with variational problems.

First, we present the rule of fractional integration by parts for the Riemann–Liouville fractional integral.

Theorem 10 Let $0 < \alpha < 1$, $p \geq 1$, $q \geq 1$ and $1/p + 1/q \leq 1 + \alpha$. If $y \in L_p$ $([a, b]; \mathbb{R})$ and $x \in L_q([a, b]; \mathbb{R})$, then the following formula for integration by parts holds:

$$\int_a^b y(t) \, {}_aI_t^\alpha x(t)dt = \int_a^b x(t) \, {}_tI_b^\alpha y(t)dt.$$

For Caputo fractional derivatives, the integration by parts formulas are presented below [3].

Theorem 11 Let $0 < \alpha < 1$. The following relations hold:

$$\int_a^b y(t) \, {}_a^C D_t^\alpha x(t)dt = \int_a^b x(t) \, {}_tD_b^\alpha y(t)dt + \left[x(t) \, {}_tI_b^{1-\alpha} y(t)\right]_{t=a}^{t=b} \quad (1.16)$$

and

$$\int_a^b y(t) \, {}_t^C D_b^\alpha x(t)dt = \int_a^b x(t) \, {}_aD_t^\alpha y(t)dt - \left[x(t) \, {}_aI_t^{1-\alpha} y(t)\right]_{t=a}^{t=b}. \quad (1.17)$$

When $\alpha \to 1$, we get ${}_a^C D_t^\alpha = {}_aD_t^\alpha = \frac{d}{dt}$, ${}_t^C D_b^\alpha = {}_tD_b^\alpha = -\frac{d}{dt}$, ${}_aI_t^\alpha = {}_tI_b^\alpha = I$, and formulas (1.16) and (1.17) give the classical formulas of integration by parts.

Then, we introduce the integration by parts formulas for variable-order fractional integrals [27].

Theorem 12 *Let $\frac{1}{n} < \alpha(t, \tau) < 1$ for all $t, \tau \in [a, b]$ and a certain $n \in \mathbb{N}$ greater or equal than two, and $x, y \in C([a, b]; \mathbb{R})$. Then the following formula for integration by parts holds:*

$$\int_a^b y(t) \, _aI_t^{\alpha(\cdot,\cdot)} x(t) dt = \int_a^b x(t) \, _tI_b^{\alpha(\cdot,\cdot)} y(t) dt.$$

In the following theorem, we present the formulas involving the Caputo fractional derivative of variable order. The theorem was proved in Odzijewicz et al. [27] and gives a generalization of the standard fractional formulas of integration by parts for a constant α.

Theorem 13 *Let $0 < \alpha(t, \tau) < 1 - \frac{1}{n}$ for all $t, \tau \in [a, b]$ and a certain $n \in \mathbb{N}$ greater or equal than two. If $x, y \in C^1([a, b]; \mathbb{R})$, then the fractional integration by parts formulas*

$$\int_a^b y(t) \, _a^C D_t^{\alpha(\cdot,\cdot)} x(t) dt = \int_a^b x(t) \, _t D_b^{\alpha(\cdot,\cdot)} y(t) dt + \left[x(t) \, _tI_b^{1-\alpha(\cdot,\cdot)} y(t) \right]_{t=a}^{t=b}$$

and

$$\int_a^b y(t) \, _t^C D_b^{\alpha(\cdot,\cdot)} x(t) dt = \int_a^b x(t) \, _a D_t^{\alpha(\cdot,\cdot)} y(t) dt - \left[x(t) \, _aI_t^{1-\alpha(\cdot,\cdot)} y(t) \right]_{t=a}^{t=b}$$

hold.

This last theorem has an important role in this work to the proof of the generalized Euler–Lagrange equations.

In the end of this chapter, we present integration by parts formulas for generalized fractional operators [19]. For that, we need the following definition:

Definition 22 Let $P = \langle a, t, b, \lambda, \mu \rangle$. We denote by P^* the parameter set $P^* = \langle a, t, b, \mu, \lambda \rangle$. The parameter P^* is called the dual of P.

Let $1 < p < \infty$ and q be the adjoint of p, that is $\frac{1}{p} + \frac{1}{q} = 1$. A proof of the next result can be found in Malinowska et al. [19].

Theorem 14 *Let $k \in L_q(\Delta; \mathbb{R})$. Then the operator K_{P^*} is a linear bounded operator from $L_p([a, b]; \mathbb{R})$ to $L_q([a, b]; \mathbb{R})$. Moreover, the following integration by parts formula holds:*

$$\int_a^b x(t) \cdot K_P[y](t) dt = \int_a^b y(t) \cdot K_{P^*}[x](t) dt$$

for all $x, y \in L_p([a, b]; \mathbb{R})$.

References

1. Abel NH (1823) Solution de quelques problèmes à l'aide d'intégrales définies. Mag Naturv 1(2):1–127
2. Agrawal OP (2010) Generalized variational problems and Euler–Lagrange equations. Comput Math Appl 59(5):1852–1864
3. Almeida R, Malinowska AB (2013) Generalized transversality conditions in fractional calculus of variations. Commun Nonlinear Sci Numer Simul 18(3):443–452
4. Almeida R, Torres DFM (2013) An expansion formula with higher-order derivatives for fractional operators of variable order. Sci World J. Art. ID 915437, 11 pp
5. Almeida R, Pooseh S, Torres DFM (2015) Computational methods in the fractional calculus of variations. Imperial College Press, London
6. Atanacković TM, Pilipovic S (2011) Hamilton's principle with variable order fractional derivatives. Fract Calc Appl Anal 14:94–109
7. Caputo M (1967) Linear model of dissipation whose Q is almost frequency independent-II. Geophys J R Astr Soc 13:529–539
8. Coimbra CFM (2003) Mechanics with variable-order differential operators. Ann Phys 12(11–12):692–703
9. Fu Z-J, Chen W, Yang H-T (2013) Boundary particle method for Laplace transformed time fractional diffusion equations. J Comput Phys 235:52–66
10. Herrmann R (2013) Folded potentials in cluster physics–a comparison of Yukawa and Coulomb potentials with Riesz fractional integrals. J Phys A 46(40):405203. 12 pp
11. Hilfer R (2000) Applications of fractional calculus in physics. World Scientific Publishing, River Edge, NJ
12. Kilbas AA, Srivastava HM, Trujillo JJ (2006) Theory and applications of fractional differential equations. Elsevier, Amsterdam
13. Klimek M (2001) Fractional sequential mechanics - models with symmetric fractional derivative. Czechoslovak J Phys 51(12):1348–1354
14. Kumar K, Pandey R, Sharma S (2017) Comparative study of three numerical schemes for fractional integro-differential equations. J Comput Appl Math 315:287–302
15. Li G, Liu H (2016) Stability analysis and synchronization for a class of fractional-order neural networks. Entropy 18:55. 13 pp
16. Li CP, Chen A, Ye J (2011) Numerical approaches to fractional calculus and fractional ordinary differential equation. J Comput Phys 230(9):3352–3368
17. Mainardi F (2010) Fractional calculus and waves in linear viscoelasticity. Imperial college press, London
18. Malinowska AB, Torres DFM (2010) Fractional variational calculus in terms of a combined Caputo derivative. In: Podlubny I, Vinagre Jara BM, Chen YQ, Feliu Batlle V, Tejado Balsera I (eds) Proceedings of FDA'10, The 4th IFAC workshop on fractional differentiation and its applications. Badajoz, Spain 18–20 Oct 2010. Article no. FDA10-084, 6 pp
19. Malinowska AB, Odzijewicz T, Torres DFM (2015) Advanced methods in the fractional calculus of variations. Springer briefs in applied sciences and technology. Springer, Cham
20. Malinowska AB, Torres DFM (2011) Fractional calculus of variations for a combined Caputo derivative. Fract Calc Appl Anal 14(4):523–537
21. Malinowska AB, Torres DFM (2012) Introduction to the fractional calculus of variations. Imperical Coll Press, London
22. Malinowska AB, Torres DFM (2012) Multiobjective fractional variational calculus in terms of a combined Caputo derivative. Appl Math Comput 218(9):5099–5111
23. Malinowska AB, Torres DFM (2012) Towards a combined fractional mechanics and quantization. Fract Calc Appl Anal 15(3):407–417
24. Odzijewicz T, Malinowska AB, Torres DFM (2012) Fractional calculus of variations in terms of a generalized fractional integral with applications to physics. Abstr Appl Anal. Art. ID 871912, 24 pp

25. Odzijewicz T, Malinowska AB, Torres DFM (2012) Fractional variational calculus with classical and combined Caputo derivatives. Nonlinear Anal 75(3):1507–1515
26. Odzijewicz T, Malinowska AB, Torres DFM (2013) Fractional variational calculus of variable order. Advances in harmonic analysis and operator theory. Operator Theory: Advances and Applications. Birkhäuser/Springer, Basel, pp 291–301
27. Odzijewicz T, Malinowska AB, Torres DFM (2013) Noether's theorem for fractional variational problems of variable order. Cent Eur J Phys 11(6):691–701
28. Odzijewicz T, Malinowska AB, Torres DFM (2013) A generalized fractional calculus of variations. Control Cybern 42(2):443–458
29. Oldham KB, Spanier J (1974) The fractional calculus. Academic Press, New York
30. Oliveira EC, Machado JAT (2014) Review of definitions for fractional derivatives and integral. A Math Probl Eng 2014:238–459 6 pp
31. Pinto C, Carvalho ARM (2014) New findings on the dynamics of HIV and TB coinfection models. Appl Math Comput 242:36–46
32. Podlubny I (1999) Fractional differential equations. Academic Press, San Diego, CA
33. Ramirez LES, Coimbra CFM (2011) On the variable order dynamics of the nonlinear wake caused by a sedimenting particle. Phys D 240(13):1111–1118
34. Ross B (1977) The development of fractional calculus 1695–1900. Historia Mathematica 4:75–89
35. Samko SG (1995) Fractional integration and differentiation of variable order. Anal Math 21(3):213–236
36. Samko SG, Ross B (1993) Integration and differentiation to a variable fractional order. Integr Transform Spec Funct 1(4):277–300
37. Samko SG, Kilbas AA, Marichev OI (1993) Fractional integrals and derivatives. Translated from the Russian original. Gordon and Breach, Yverdon (1987)
38. Sheng H, Sun HG, Coopmans C, Chen YQ, Bohannan GW (2011) A physical experimental study of variable-order fractional integrator and differentiator. Eur Phys J 193(1):93–104
39. Sierociuk D, Skovranek T, Macias M, Podlubny I, Petras I, Dzielinski A, Ziubinski P (2015) Diffusion process modeling by using fractional-order models. Appl Math Comput 257(15):2–11
40. Sun HG, Chen W, Chen YQ (2009) Variable order fractional differential operators in anomalous diffusion modeling. Phys A 388(21):4586–4592
41. Sun H, Chen W, Li C, Chen Y (2012) Finite difference schemes for variable-order time fractional diffusion equation. Int J Bifur Chaos Appl Sci Eng 22(4):1250085. 16 pp
42. Sun H, Hu S, Chen Y, Chen W, Yu Z (2013) A dynamic-order fractional dynamic system. Chin Phys Lett 30(4):046601. 4 pp

Chapter 2
The Calculus of Variations

As part of this book is devoted to the fractional calculus of variations, in this chapter, we introduce the basic concepts about the classical calculus of variations and the fractional calculus of variations. The study of fractional problems of the calculus of variations and respective Euler–Lagrange-type equations is a subject of current strong research.

In Sect. 2.1, we introduce some concepts and important results from the classical theory. Afterward, in Sect. 2.2, we start with a brief historical introduction to the noninteger calculus of variations, and then, we present recent results on the fractional calculus of variations.

For more information about this subject, we refer the reader to the books Almeida et al. [7], Malinowska et al. [23], Malinowska and Torres [22], and van Brunt [36].

2.1 The Classical Calculus of Variations

The calculus of variations is a field of mathematical analysis that concerns with finding extrema (maxima or minima) for functionals, i.e., concerns with the problem of finding a function for which the value of a certain integral is either the largest or the smallest possible.

In this context, a functional is a mapping from a set of functions to the real numbers; i.e., it receives a function and produces a real number. Let $D \subseteq C^2([a, b]; \mathbb{R})$ be a linear space endowed with a norm $\| \cdot \|$. The cost functional $\mathcal{J} : D \to \mathbb{R}$ is generally of the form

$$\mathcal{J}(x) = \int_a^b L\left(t, x(t), x'(t)\right) dt, \tag{2.1}$$

where $t \in [a, b]$ is the independent variable, usually called time, and $x(t) \in \mathbb{R}$ is a function. The integrand $L : [a, b] \times \mathbb{R}^2 \to \mathbb{R}$, that depends on the function x, its

R. Almeida et al., *The Variable-Order Fractional Calculus of Variations*, SpringerBriefs in Applied Sciences and Technology, https://doi.org/10.1007/978-3-319-94006-9_2

derivative x', and the independent variable t, is a real-valued function, called the Lagrangian.

The roots of the calculus of variations appear in works of Greek thinkers, such as Queen Dido or Aristotle in the late of the first century BC. During the seventeenth century, some physicists and mathematicians (Galileo, Fermat, Newton, among others) investigated some variational problems, but in general they did not use variational methods to solve them. The development of the calculus of variations began with a problem posed by Johann Bernoulli in 1696, called the brachistochrone problem: given two points A and B in a vertical plane, what is the curve traced out by a point acted on only by gravity, which starts at A and reaches B in minimal time? The curve that solves the problem is called the *brachistochrone*. This problem caught the attention of some mathematicians including Jakob Bernoulli, Leibniz, L'Hôpital, and Newton, which presented also a solution for the brachistochrone problem. Integer variational calculus is still, nowadays, a relevant area of research. It plays a significant role in many areas of science, physics, engineering, economics, and applied mathematics.

The classical variational problem, considered by Leonhard Euler, is stated as follows.

Among all functions $x \in D$ with $a, b \in \mathbb{R}$, find the ones that minimize (or maximize) the functional $\mathcal{J} : D \to \mathbb{R}$, with

$$\mathcal{J}(x) = \int_a^b L\left(t, x(t), x'(t)\right) dt \tag{2.2}$$

subject to the boundary conditions

$$x(a) = x_a, \qquad x(b) = x_b, \tag{2.3}$$

with x_a, x_b fixed reals and the Lagrangian L satisfying some smoothness properties. Usually, we say that a function is "sufficiently smooth" for a particular development if all required actions (integration, differentiation, etc.) are possible.

Definition 23 A trajectory $x \in C^2([a, b]; \mathbb{R})$ is said to be an admissible trajectory if it satisfies all the constraints of the problem along the interval $[a, b]$. The set of admissible trajectories is denoted by D.

To discuss maxima and minima of functionals, we need to introduce the following definition.

Definition 24 We say that $x^* \in D$ is a local extremizer to the functional $\mathcal{J} : D \to \mathbb{R}$ if there exists some real $\epsilon > 0$, such that

$$\forall x \in D : \quad \|x^* - x\| < \epsilon \implies \mathcal{J}(x^*) - \mathcal{J}(x) \leq 0 \vee \mathcal{J}(x^*) - \mathcal{J}(x) \geq 0.$$

In this context, as we are dealing with functionals defined on functions, we need to clarify the term of directional derivatives, here called variations. The concept of variation of a functional is central to obtain the solution of variational problems.

Definition 25 Let $\mathcal{J}$ be defined on D. The first variation of a functional $\mathcal{J}$ at $x \in D$ in the direction $h \in D$ is defined by

$$\delta \mathcal{J}(x, h) = \lim_{\epsilon \to 0} \frac{\mathcal{J}(x + \epsilon h) - \mathcal{J}(x)}{\epsilon} = \frac{d}{d\epsilon} \mathcal{J}(x + \epsilon h) \bigg]_{\epsilon=0},$$

where x and h are functions and ϵ is a scalar, whenever the limit exists.

Definition 26 A direction $h \in D$, $h \neq 0$, is said to be an admissible variation for $\mathcal{J}$ at $y \in D$ if

1. $\delta \mathcal{J}(x, h)$ exists;
2. $x + \epsilon h \in D$ for all sufficiently small ϵ.

With the condition that $\mathcal{J}(x)$ be a local extremum and the definition of variation, we have the following result that offers a necessary optimality condition for problems of calculus of variations [36].

Theorem 15 *Let $\mathcal{J}$ be a functional defined on D. If x^* minimizes (or maximizes) the functional $\mathcal{J}$ over all functions $x : [a, b] \to \mathbb{R}$ satisfying boundary conditions (2.3), then*

$$\delta \mathcal{J}(x^*, h) = 0$$

for all admissible variations h at x^*.

2.1.1 Euler–Lagrange Equations

Although the calculus of variations was born with Johann's problem, it was with the work of Euler in 1742 and the one of Lagrange in 1755 that a systematic theory was developed. The common procedure to address such variational problems consists in solving a differential equation, called the Euler–Lagrange equation, which every minimizer/maximizer of the functional must satisfy.

In Lemma 16, we review an important result to transform the necessary condition of extremum in a differential equation, free of integration with an arbitrary function. In the literature, it is known as the fundamental lemma of the calculus of variations.

Lemma 16 *Let x be continuous in $[a, b]$ and let h be an arbitrary function on $[a, b]$ such that it is continuous and $h(a) = h(b) = 0$. If*

$$\int_a^b x(t)h(t)dt = 0$$

for all such h, then $x(t) = 0$ for all $t \in [a, b]$.

For the sequel, we denote by $\partial_i z$, $i \in \{1, 2, \ldots, M\}$, with $M \in \mathbb{N}$, the partial derivative of a function $z : \mathbb{R}^M \to \mathbb{R}$ with respect to its ith argument. Now we can formulate the necessary optimality condition for the classical variational problem [36].

Theorem 17 *If x is an extremizing of the functional (2.2) on D, subject to (2.3), then x satisfies*

$$\partial_2 L \left(t, x(t), x'(t) \right) - \frac{d}{dt} \partial_3 L \left(t, x(t), x'(t) \right) = 0 \qquad (2.4)$$

for all $t \in [a, b]$.

To solve this second-order differential equation, the two given boundary conditions (2.3) provide sufficient information to determine the two arbitrary constants.

Definition 27 A curve x that is a solution of the Euler–Lagrange differential equation will be called an extremal of $\mathcal{J}$.

2.1.2 Problems with Variable Endpoints

In the basic variational problem considered previously, the functional $\mathcal{J}$ to minimize (or maximize) is subject to given boundary conditions of the form

$$x(a) = x_a, \qquad x(b) = x_b,$$

where $x_a, x_b \in \mathbb{R}$ are fixed. It means that the solution of the problem, x, needs to pass through the prescribed points. This variational problem is called fixed endpoints variational problem. The Euler–Lagrange equation (2.4) is normally a second-order differential equation containing two arbitrary constants, so with two given boundary conditions provided, they are sufficient to determine the two constants.

However, in some areas, like physics and geometry, the variational problems do not impose the appropriate number of boundary conditions. In these cases, when one or both boundary conditions are missing, that is, when the set of admissible functions may take any value at one or both of the boundaries, then one or two auxiliary conditions, known as the natural boundary conditions or transversality conditions, need to be obtained in order to solve the equation [36]:

$$\left[\frac{\partial L(t, x(t), x'(t))}{\partial x'} \right]_{t=a} = 0 \quad \text{and/or} \quad \left[\frac{\partial L(t, x(t), x'(t))}{\partial x'} \right]_{t=b} = 0. \qquad (2.5)$$

There are different types of variational problems with variable endpoints:

- Free terminal point–one boundary condition at the initial time ($x(a) = x_a$). The terminal point is free ($x(b) \in \mathbb{R}$);

- Free initial point–one boundary condition at the final time ($x(b) = x_b$). The initial point is free ($x(a) \in \mathbb{R}$);
- Free endpoints–both endpoints are free ($x(a) \in \mathbb{R}$, $x(b) \in \mathbb{R}$);
- Variable endpoints–the initial point $x(a)$ or/and the endpoint $x(b)$ is variable on a certain set, for example, on a prescribed curve.

Another generalization of the variational problem consists to find an optimal curve x and the optimal final time T of the variational integral, $T \in [a, b]$. This problem is known in the literature as a free-time problem [15]. An example is the following free-time problem with free terminal point. Let D denote the subset $C^2([a, b]; \mathbb{R}) \times [a, b]$ endowed with a norm $\|(\cdot, \cdot)\|$. Find the local minimizers of the functional $\mathcal{J} : D \rightarrow \mathbb{R}$, with

$$\mathcal{J}(x, T) = \int_a^T L(t, x(t), x'(t))dt, \tag{2.6}$$

over all $(x, T) \in D$ satisfying the boundary condition $x(a) = x_a$, with $x_a \in \mathbb{R}$ fixed. The terminal time T and the terminal state $x(T)$ are here both free.

Definition 28 We say that $(x^\star, T^\star) \in D$ is a local extremizer (minimizer or maximizer) to the functional $\mathcal{J} : D \rightarrow \mathbb{R}$ as in (2.6) if there exists some $\epsilon > 0$ such that, for all $(x, T) \in D$,

$$\|(x^\star, T^\star) - (x, T)\| < \epsilon \Rightarrow J(x^\star, T^\star) \leq J(x, T) \vee J(x^\star, T^\star) \geq J(x, T).$$

To develop a necessary optimality condition to problem (2.6) for an extremizer $(x^\star, T^\star)$, we need to consider an admissible variation of the form:

$$(x^\star + \epsilon h, \quad T^\star + \epsilon \Delta T),$$

where $h \in C^1([a, b]; \mathbb{R})$ is a perturbing curve that satisfies the condition $h(a) = 0$, ϵ represents a small real number, and ΔT represents an arbitrarily chosen small change in T. Considering the functional $\mathcal{J}(x, T)$ in this admissible variation, we get a function of ϵ, where the upper limit of integration will also vary with ϵ:

$$\mathcal{J}(x^\star + \epsilon h, \quad T^\star + \epsilon \Delta T) = \int_a^{T^\star + \epsilon \Delta T} L(t, (x^\star + \epsilon h)(t), (x^\star + \epsilon h)'(t))dt. \tag{2.7}$$

To find the first-order necessary optimality condition, we need to determine the derivative of (2.7) with respect to ϵ and set it equal to zero. By doing it, we obtain three terms on the equation, where the Euler–Lagrange equation emerges from the first term, and the other two terms, which depend only on the terminal time T, give the transversality conditions.

2.1.3 Constrained Variational Problems

Variational problems are often subject to one or more constraints (holonomic constraints, integral constraints, dynamic constraints, etc.). Isoperimetric problems are a special class of constrained variational problems for which the admissible functions are needed to satisfy an integral constraint.

Here, we review the classical isoperimetric variational problem. The classical variational problem, already defined, may be modified by demanding that the class of potential extremizing functions also satisfies a new condition, called an isoperimetric constraint, of the form

$$\int_a^b g(t, x(t), x'(t))dt = C, \tag{2.8}$$

where g is a given function of t, x and x', and C is a given real number.

The new problem is called an isoperimetric problem and encompasses an important family of variational problems. In this case, the variational problems are often subject to one or more constraints involving an integral of a given function [18]. Some classical examples of isoperimetric problems appear in geometry. The most famous example consists in finding the curve of a given perimeter that bounds the greatest area, and the answer is the circle. Isoperimetric problems are an important type of variational problems, with applications in different areas, like geometry, astronomy, physics, algebra, or analysis.

In the next theorem, we present a necessary condition for a function to be an extremizer to a classical isoperimetric problem, obtained via the concept of Lagrange multiplier [36].

Theorem 18 *Consider the problem of minimizing (or maximizing) the functional $\mathcal{J}$, defined by (2.2), on D given by those $x \in C^2\left([a, b]; \mathbb{R}\right)$ satisfying the boundary conditions (2.3) and an integral constraint of the form*

$$\mathcal{G} = \int_a^b g(t, x(t), x'(t))dt = C,$$

where $g : [a, b] \times \mathbb{R}^2 \to \mathbb{R}$ is a twice continuously differentiable function. Suppose that x gives a local minimum (or maximum) to this problem. Assume that $\delta \mathcal{G}(x, h)$ does not vanish for all $h \in D$. Then, there exists a constant λ such that x satisfies the Euler–Lagrange equation

$$\partial_2 F\left(t, x(t), x'(t), \lambda\right) - \frac{d}{dt}\partial_3 F\left(t, x(t), x'(t), \lambda\right) = 0, \tag{2.9}$$

where $F\left(t, x, x', \lambda\right) = L(t, x, x') - \lambda g(t, x, x')$.

Remark 19 The constant λ is called a Lagrange multiplier.

Observe that $\delta\mathcal{G}(x, h)$ does not vanish for all $h \in D$ if x does not satisfy the Euler–Lagrange equation with respect to the isoperimetric constraint; that is, x is not an extremal for $\mathcal{G}$.

2.2 Fractional Calculus of Variations

The first connection between fractional calculus and the calculus of variations appeared in the XIX century, with Niels Abel [1]. In 1823, Abel applied fractional calculus in the solution of an integral equation involved in a generalization of the tautochrone problem. Only in the XX century, however, both areas were joined in an unique research field: the fractional calculus of variations.

The fractional calculus of variations deals with problems in which the functional, the constraint conditions, or both, depend on some fractional operator [7, 22, 23], and the main goal is to find functions that extremize such a fractional functional. By inserting fractional operators that are nonlocal in variational problems, they are suitable for developing some models possessing memory effects.

This is a fast-growing subject, and different approaches have been developed by considering different types of Lagrangians, e.g., depending on Riemann–Liouville or Caputo fractional derivatives, fractional integrals, and mixed integer-fractional-order operators (see, e.g., Almeida and Torres [6], Askari and Ansari [8], Atanacković et al. [10], Baleanu [12], Baleanu et al. [13], Cresson [17], Rabei et al. [27], Tarasov [34]). In recent years, there has been a growing interest in the area of fractional variational calculus and its applications, which include classical and quantum mechanics, field theory, and optimal control.

Although the origin of fractional calculus goes back more than three centuries, the calculus of variations with fractional derivatives has born with the works of F. Riewe only in 1996–1997. In his works, Riewe obtained a version of the Euler–Lagrange equations for problems of the calculus of variations with fractional derivatives, when investigating nonconservative Lagrangian and Hamiltonian mechanics [28, 29]. Agrawal continued the study of the fractional Euler–Lagrange equations [2–4], for some kinds of fractional variational problems, for example, problems with only one dependent variable, for functionals with different orders of fractional derivatives, for several functions, involving both Riemann–Liouville and Caputo derivatives, etc. The most common fractional operators considered in the literature take into account the past of the process, that is, one usually uses left fractional operators. But, in some cases, we may be also interested in the future of the process, and the computation of $\alpha(\cdot)$ to be influenced by it. In that case, right fractional derivatives are then considered.

2.2.1 Fractional Euler–Lagrange Equations

Similarly to the classical variational calculus, the common procedure to address such fractional variational problems consists in solving a fractional differential equation, called the fractional Euler–Lagrange equation, which every minimizer/maximizer of the functional must satisfy. With the help of the boundary conditions imposed on the problem at the initial time $t = a$ and at the terminal time $t = b$, one solves, often with the help of some numerical procedure, the fractional differential equation and obtain a possible solution to the problem [14, 19, 26, 33, 37].

Referring again to Riewe's works [28, 29], friction forces are described with Lagrangians that contain fractional derivatives. Precisely, for r, N, and N' natural numbers and assuming $x : [a, b] \to \mathbb{R}^r$, $\alpha_i, \beta_j \in [0, 1]$ with $i = 1, \ldots, N, j = 1, \ldots, N'$, the functional defined by Riewe is

$$\mathcal{J}(x) = \tag{2.10}$$

$$\int_a^b L\left({_aD_t^{\alpha_1}}[x](t), \ldots, {_aD_t^{\alpha_N}}[x](t), {_tD_b^{\beta_1}}[x](t), \ldots, {_tD_b^{\beta_{N'}}}[x](t), x(t), t \right) dt.$$

He proved that any solution x of the variational problem of extremizing the functional (2.10) satisfies the following necessary condition:

Theorem 20 *Considering the variational problem of minimizing (or maximizing) the functional* (2.10), *the fractional Euler–Lagrange equation is*

$$\sum_{i=1}^{N} {_tD_b^{\alpha_i}}[\partial_i L] + \sum_{i=1}^{N'} {_aD_t^{\beta_i}}[\partial_{i+N} L] + \partial_{N'+N+1} L = 0.$$

Riewe also illustrated his results considering the classical problem of linear friction [28].

In what follows, we are concerned with problems of the fractional calculus of variations where the functional depends on a combined fractional Caputo derivative with constant orders α and β [21, Definition 15].

Let D denote the set of all functions $x : [a, b] \to \mathbb{R}^N$, endowed with a norm $\| \cdot \|$ in $\mathbb{R}^N$. Consider the following problem: Find a function $x \in D$ for which the functional

$$\mathcal{J}(x) = \int_a^b L(t, x(t), {^CD_\gamma^{\alpha,\beta}}x(t)) dt \tag{2.11}$$

subject to given boundary conditions

$$x(a) = x_a, \qquad x(b) = x_b$$

archives a minimum, where $t \in [a, b]$, $x_a, x_b \in \mathbb{R}^N$, $\gamma \in [0, 1]$ and the Lagrangian L satisfies some smoothness properties.

Theorem 21 *Let $x = (x_1, \ldots, x_N)$ be a local minimizer to the problem with the functional (2.11) subject to two boundary conditions, as defined before. Then, x satisfies the system of N fractional Euler–Lagrange equations*

$$\partial_i L(t, x(t), {}^C D_\gamma^{\alpha,\beta} x(t)) + D_{1-\gamma}^{\beta,\alpha} \partial_{N+i} L(t, x(t), {}^C D_\gamma^{\alpha,\beta} x(t)) = 0, \qquad (2.12)$$

$i = 2, \ldots, N + 1$, *for all* $t \in [a, b]$.

For a proof of the last result, see Malinowska and Torres [20]. Observe that, if the orders α and β go to 1, and if $\gamma = 0$ or $\gamma = 1$, we obtain a corresponding result in the classical context of the calculus of variations. In fact, considering α and β going to 1, the fractional derivatives ${}_a^C D_t^\alpha$ and ${}_a D_t^\alpha$ coincide with the classical derivative $\frac{d}{dt}$; and similarly, ${}_t^C D_b^\beta$ and ${}_t D_b^\beta$ coincide with the classical derivative $-\frac{d}{dt}$.

Variational problems with free endpoints and transversality conditions are also relevant subjects in the fractional calculus of variations. The subject of free boundary points in fractional variational problems was first considered by Agrawal in [3]. In that work, he studied the Euler–Lagrange equation and transversality conditions for the case when both initial and final times are given and the admissible functions are specified at the initial time but are unspecified at the final time. After that, some free-time variational problems involving fractional derivatives or/and fractional integrals were studied [5, 24].

Sometimes, the analytic solution of the fractional Euler–Lagrange equation is very difficult to obtain, and, in this case, some numerical methods have been developed to solve the variational problem [7].

2.2.2 Fractional Variational Problems of Variable-Order

In recent years, motivated by the works of Samko and Ross, where they investigated integrals and derivatives not of a constant but of variable-order [30, 31], some problems of the calculus of variations involving derivatives of variable fractional order have appeared [9, 25].

Considering Definition 19 of the generalized fractional integral of operator K_P, Malinowska et al. [23] presented a new variational problem, where the functional was defined by a given kernel. For appropriate choices of the kernel k and the set P, we can obtain a variable-order fractional variational problem (see Remark 9).

Let $P = \langle a, t, b, \lambda, \mu \rangle$. Consider the functional $\mathcal{J}$ in $\mathbf{A}(x_a, x_b)$ defined by:

$$\mathcal{J}[x] = \int_a^b L\left(x(t), K_P[x](t), x'(t), B_P[x](t), t\right) dt, \qquad (2.13)$$

where $\mathbf{A}(x_a, x_b)$ is the set

$$\{x \in C^1([a, b]; \mathbb{R}) : x(a) = x_a, x(b) = x_b, K_P[x], B_P[x] \in C([a, b]; \mathbb{R})\},$$

and K_P is the generalized fractional integral operator with kernel belonging to $L_q(\Delta; \mathbb{R})$ and B_P the generalized fractional derivative of Caputo type.

The optimality condition for the problem that consists to determine a function that minimize (or maximize) the functional (2.13) is given in the following theorem [23].

Theorem 22 *Let $x \in \mathbf{A}(x_a, x_b)$ be a minimizer of functional (2.13). Then, x satisfies the following Euler–Lagrange equation:*

$$\frac{d}{dt}[\partial_3 L(\star_x)(t)] + A_{P^*}[\tau \longmapsto \partial_4 L(\star_x)(\tau)](t)$$
$$= \partial_1 L(\star_x)(t) + K_{P^*}[\tau \longmapsto \partial_2 L(\star_x)(\tau)](t), \tag{2.14}$$

where $(\star_x)(t) = (x(t), K_P[x](t), x'(t), B_P[x](t), t)$, for $t \in (a, b)$.

Observe that if functional (2.13) does not depends on the generalized fractional operators K_P and B_P, this problem coincides with the classical variational problem and Theorem 22 reduces to Theorem 17.

Let $\Delta := \{(t, \tau) \in \mathbb{R}^2 : a \leq \tau < t \leq b\}$ and let $1 < p < \infty$ and q be the adjoint of p. A special case of this problem is obtained when we consider $\alpha : \Delta \to [0, 1 - \delta]$ with $\delta > 1/p$ and the kernel is defined by

$$k^\alpha(t, \tau) = \frac{1}{\Gamma(1 - \alpha(t, \tau))}(t - \tau)^{-\alpha(t, \tau)}$$

in $L_q(\Delta; \mathbb{R})$. The next results provides necessary conditions of optimality [23].

Theorem 23 *Consider the problem of minimizing a functional*

$$\mathcal{J}[x] = \int_a^b L\left(x(t), {}_a I_t^{1-\alpha(\cdot,\cdot)}[x](t), x'(t), {}_a^C D_t^{\alpha(\cdot,\cdot)}[x](t), t\right) dt \tag{2.15}$$

subject to boundary conditions

$$x(a) = x_a, \quad x(b) = x_b, \tag{2.16}$$

where $x', {}_a I_t^{1-\alpha(\cdot,\cdot)}[x], {}_a^C D_t^{\alpha(\cdot,\cdot)}[x] \in C([a, b]; \mathbb{R})$. Then, if $x \in C^1([a, b]; \mathbb{R})$ minimizes (or maximizes) the functional (2.15) subject to (2.16), then it satisfies the following Euler–Lagrange equation:

$$\partial_1 L\left(x(t), {}_a I_t^{1-\alpha(\cdot,\cdot)}[x](t), x'(t), {}_a^C D_t^{\alpha(\cdot,\cdot)}[x](t), t\right)$$
$$- \frac{d}{dt}\partial_3 L\left(x(t), {}_a I_t^{1-\alpha(\cdot,\cdot)}[x](t), x'(t), {}_a^C D_t^{\alpha(\cdot,\cdot)}[x](t), t\right)$$
$$+ {}_t I_b^{1-\alpha(\cdot,\cdot)}\left[\partial_2 L\left(x(\tau), {}_a I_\tau^{1-\alpha(\cdot,\cdot)}[x](\tau), x'(\tau), {}_a^C D_\tau^{\alpha(\cdot,\cdot)}[x](\tau), \tau\right)\right](t)$$
$$+ {}_t D_b^{\alpha(\cdot,\cdot)}\left[\partial_4 L\left(x(\tau), {}_a I_\tau^{1-\alpha(\cdot,\cdot)}[x](\tau), x'(\tau), {}_a^C D_\tau^{\alpha(\cdot,\cdot)}[x](\tau), \tau\right)\right](t) = 0.$$

In fact, the use of fractional derivatives of constant order in variational problems may not be the best option, since trajectories are a dynamic process, and the order may vary. Therefore, it is important to consider the order to be a function, $\alpha(\cdot)$, depending on time. Then, we may seek what is the best function $\alpha(\cdot)$ such that the variable-order fractional differential equation $D^{\alpha(\cdot)}x(t) = f(t, x(t))$ better describes the process under study.

This approach is very recent, and many works have to be done for a complete study of the subject (see, e.g., Atangana and Kilicman [11], Coimbra et al. [16], Samko and Ross [31], Sheng et al. [32], Valério et al. [35]).

References

1. Abel NH (1823) Solution de quelques problèmes à l'aide d'intégrales définies. Mag Naturv 1(2):1–127
2. Agrawal OP (2002) Formulation of Euler–Lagrange equations for fractional variational problems. J Math Anal Appl 272:368–379
3. Agrawal OP (2006) Fractional variational calculus and the transversality conditions. J Phys A: Math Gen 39(33):10375–10384
4. Agrawal OP (2007) Generalized Euler-Lagrange equations and transversality conditions for FVPs in terms of the Caputo derivative. J Vib Control 13(9–10):1217–1237
5. Almeida R, Malinowska AB (2013) Generalized transversality conditions in fractional calculus of variations. Commun Nonlinear Sci Numer Simul 18(3):443–452
6. Almeida R, Torres DFM (2009) Calculus of variations with fractional derivatives and fractional integrals. Appl Math Lett 22(12):1816–1820
7. Almeida R, Pooseh S, Torres DFM (2015) Computational methods in the fractional calculus of variations. Imperial College Press, London
8. Askari H, Ansari A (2016) Fractional calculus of variations with a generalized fractional derivative. Fract Differ Calc 6:57–72
9. Atanacković TM, Pilipovic S (2011) Hamilton's principle with variable order fractional derivatives. Fract Calc Appl Anal 14:94–109
10. Atanacković TM, Konjik S, Pilipović S (2008) Variational problems with fractional derivatives: Euler–Lagrange equations. J Phys A 41(9):095201 12 pp
11. Atangana A, Kilicman A (2014) On the generalized mass transport equation to the concept of variable fractional derivative. Math Probl Eng 2014. Art. ID 542809, 9 pp
12. Baleanu D (2008) New applications of fractional variational principles. Rep Math Phys 61(2):199–206
13. Baleanu D, Golmankhaneh AK, Nigmatullin R, Golmankhaneh AK (2010) Fractional Newtonian mechanics. Cent Eur J Phys 8(1):120–125
14. Blaszczyk T, Ciesielski M (2014) Numerical solution of fractional Sturm-Liouville equation in integral form. Fract Calc Appl Anal 17(2):307–320
15. Chiang AC (1992) Elements of dynamic optimization. McGraw-Hill Inc, Singapore
16. Coimbra CFM, Soon CM, Kobayashi MH (2005) The variable viscoelasticity operator. Annalen der Physik 14(6):378–389
17. Cresson J (2007) Fractional embedding of differential operators and Lagrangian systems. J Math Phys 48(3):033504 34 pp
18. Fraser C (1992) Isoperimetric problems in variatonal calculus of Euler and Lagrange. Historia Mathematica 19:4–23
19. Lotfi A, Yousefi SA (2013) A numerical technique for solving a class of fractional variational problems. J Comput Appl Math 237(1):633–643

20. Malinowska AB, Torres DFM (2010) Fractional variational calculus in terms of a combined Caputo derivative. In: : Podlubny I, Vinagre Jara BM, Chen YQ, Feliu Batlle V, Tejado Balsera I (eds) Proceedings of FDA'10, The 4th IFAC workshop on fractional differentiation and its applications. Badajoz, Spain, 18–20 Oct 2010. Article no. FDA10-084, 6 pp
21. Malinowska AB, Torres DFM (2011) Fractional calculus of variations for a combined Caputo derivative. Fract Calc Appl Anal 14(4):523–537
22. Malinowska AB, Torres DFM (2012) Introduction to the fractional calculus of variations. Imperial Press, London
23. Malinowska AB, Odzijewicz T, Torres DFM (2015) Advanced methods in the fractional calculus of variations. Springer briefs in applied sciences and technology. Springer, Cham
24. Odzijewicz T, Malinowska AB, Torres DFM (2012) Fractional variational calculus with classical and combined Caputo derivatives. Nonlinear Anal 75(3):1507–1515
25. Odzijewicz T, Malinowska AB, Torres DFM (2012) Variable order fractional variational calculus for double integrals. In: Proceedings of the 51st IEEE conference on decision and control, 10–13 Dec 2012. Maui, Hawaii, Art. no. 6426489, pp 6873–6878
26. Pooseh S, Almeida R, Torres DFM (2013) Discrete direct methods in the fractional calculus of variations. Comput Math Appl 66(5):668–676
27. Rabei EM, Nawafleh KI, Hijjawi RS, Muslih SI, Baleanu D (2007) The Hamilton formalism with fractional derivatives. J Math Anal Appl 327(2):891–897
28. Riewe F (1996) Nonconservative Lagrangian and Hamiltonian mechanics. Phys Rev E (3) 53(2):1890–1899
29. Riewe F (1997) Mechanics with fractional derivatives. Phys Rev E (3) 55(3):3581–3592
30. Samko SG (1995) Fractional integration and differentiation of variable order. Anal Math 21(3):213–236
31. Samko SG, Ross B (1993) Integration and differentiation to a variable fractional order. Integr Transform Spec Funct 1(4):277–300
32. Sheng H, Sun HG, Coopmans C, Chen YQ, Bohannan GW (2011) A physical experimental study of variable-order fractional integrator and differentiator. Eur Phys J 193(1):93–104
33. Sumelka W, Blaszczyk T (2014) Fractional continua for linear elasticity. Arch Mech 66(3):147–172
34. Tarasov VE (2006) Fractional variations for dynamical systems: Hamilton and Lagrange approaches. J Phys A 39(26):8409–8425
35. Valério D, Vinagre G, Domingues J, Costa JS (2009) Variable-order fractional derivatives and their numerical approximations I – real orders. In: Ortigueira et al (eds) Symposium on fractional signals and systems Lisbon 09. Lisbon, Portugal
36. van Brunt B (2004) The calculus of variations. Universitext. Springer, New York
37. Xu Y, Agrawal OP (2014) Models and numerical solutions of generalized oscillator equations. J Vib Acoust 136(5):050903 7 pp

Chapter 3
Expansion Formulas for Fractional Derivatives

In this chapter, we present a new numerical tool to solve differential equations involving three types of Caputo derivatives of fractional variable order. For each one of them, an approximation formula is obtained, which is written in terms of standard (integer order) derivatives only. Estimations for the error of the approximations are also provided. Then, we compare the numerical approximation of some test function with its exact fractional derivative. We present an exemplification of how the presented methods can be used to solve partial fractional differential equations of variable-order.

Let us briefly describe the main contents of the chapter. We begin this chapter by formulating the needed definitions (Sect. 3.1), namely, we present three types of Caputo derivatives of variable fractional order. First, we consider one independent variable only; then, we generalize for several independent variables. The following Sect. 3.2 is the main core of the chapter, where we prove approximation formulas for the given fractional operators of variable order and, respectively, upper-bound formulas for the errors. To test the efficiency of the proposed method, in Sect. 3.3, we compare the exact fractional derivative of some test function with the numerical approximations obtained from the decomposition formulas given in Sect. 3.2. To end, in Sect. 3.4, we apply our method to approximate two physical problems involving Caputo fractional operators of variable-order (a time-fractional diffusion equation and a fractional Burgers' partial differential equation in fluid mechanics) by classical problems that may be solved by well-known standard techniques.

The results of this chapter first appeared in Tavares et al. [17].

3.1 Caputo-Type Fractional Operators of Variable-Order

In the literature of fractional calculus, several different definitions of derivatives are found [12]. One of those, introduced by Caputo [1] and studied independently by

R. Almeida et al., *The Variable-Order Fractional Calculus of Variations*, SpringerBriefs in Applied Sciences and Technology, https://doi.org/10.1007/978-3-319-94006-9_3

other authors, like Džrbašjan and Nersesjan [5] and Rabotnov [10], has found many applications and seems to be more suitable to model physical phenomena [3, 4, 7, 8, 14, 16, 18].

3.1.1 Caputo Derivatives for Functions of One Variable

Our goal is to consider fractional derivatives of variable-order, with α depending on time. In fact, some phenomena in physics are better described when the order of the fractional operator is not constant, for example, in the diffusion process in an inhomogeneous or heterogeneous medium, or processes where the changes in the environment modify the dynamic of the particle [2, 13, 15]. Motivated by the above considerations, we introduce three types of Caputo fractional derivatives. The order of the derivative is considered as a function $\alpha(t)$ taking values on the open interval $(0, 1)$. To start, we define two different kinds of Riemann–Liouville fractional derivatives.

Definition 29 Given a function $x : [a, b] \to \mathbb{R}$,

1. the type I left Riemann–Liouville fractional derivative of order $\alpha(t)$ is defined by

$$_aD_t^{\alpha(t)}x(t) = \frac{1}{\Gamma(1 - \alpha(t))} \frac{d}{dt} \int_a^t (t - \tau)^{-\alpha(t)} x(\tau) d\tau;$$

2. the type I right Riemann–Liouville fractional derivative of order $\alpha(t)$ is defined by

$$_tD_b^{\alpha(t)}x(t) = \frac{-1}{\Gamma(1 - \alpha(t))} \frac{d}{dt} \int_t^b (\tau - t)^{-\alpha(t)} x(\tau) d\tau;$$

3. the type II left Riemann–Liouville fractional derivative of order $\alpha(t)$ is defined by

$$_a\mathcal{D}_t^{\alpha(t)}x(t) = \frac{d}{dt} \left(\frac{1}{\Gamma(1 - \alpha(t))} \int_a^t (t - \tau)^{-\alpha(t)} x(\tau) d\tau \right);$$

4. the type II right Riemann–Liouville fractional derivative of order $\alpha(t)$ is defined by

$$_t\mathcal{D}_b^{\alpha(t)}x(t) = \frac{d}{dt} \left(\frac{-1}{\Gamma(1 - \alpha(t))} \int_t^b (\tau - t)^{-\alpha(t)} x(\tau) d\tau \right).$$

The Caputo derivatives are given using the previous Riemann–Liouville fractional derivatives.

Definition 30 Given a function $x : [a, b] \to \mathbb{R}$,

1. the type I left Caputo derivative of order $\alpha(t)$ is defined by

$$\substack{C \\ a}D_t^{\alpha(t)}x(t) = {}_aD_t^{\alpha(t)}(x(t) - x(a))$$

$$= \frac{1}{\Gamma(1 - \alpha(t))} \frac{d}{dt} \int_a^t (t - \tau)^{-\alpha(t)}[x(\tau) - x(a)]d\tau;$$

2. the type I right Caputo derivative of order $\alpha(t)$ is defined by

$$\substack{C \\ t}D_b^{\alpha(t)}x(t) = {}_tD_b^{\alpha(t)}(x(t) - x(b))$$

$$= \frac{-1}{\Gamma(1 - \alpha(t))} \frac{d}{dt} \int_t^b (\tau - t)^{-\alpha(t)}[x(\tau) - x(b)]d\tau;$$

3. the type II left Caputo derivative of order $\alpha(t)$ is defined by

$$\substack{C \\ a}\mathcal{D}_t^{\alpha(t)}x(t) = {}_a\mathcal{D}_t^{\alpha(t)}(x(t) - x(a))$$

$$= \frac{d}{dt}\left(\frac{1}{\Gamma(1 - \alpha(t))} \int_a^t (t - \tau)^{-\alpha(t)}[x(\tau) - x(a)]d\tau\right);$$

4. the type II right Caputo derivative of order $\alpha(t)$ is defined by

$$\substack{C \\ t}\mathcal{D}_b^{\alpha(t)}x(t) = {}_t\mathcal{D}_b^{\alpha(t)}(x(t) - x(b))$$

$$= \frac{d}{dt}\left(\frac{-1}{\Gamma(1 - \alpha(t))} \int_t^b (\tau - t)^{-\alpha(t)}[x(\tau) - x(b)]d\tau\right);$$

5. the type III left Caputo derivative of order $\alpha(t)$ is defined by

$$\substack{C \\ a}\mathbb{D}_t^{\alpha(t)}x(t) = \frac{1}{\Gamma(1 - \alpha(t))} \int_a^t (t - \tau)^{-\alpha(t)}x'(\tau)d\tau;$$

6. the type III right Caputo derivative of order $\alpha(t)$ is defined by

$$\substack{C \\ t}\mathbb{D}_b^{\alpha(t)}x(t) = \frac{-1}{\Gamma(1 - \alpha(t))} \int_t^b (\tau - t)^{-\alpha(t)}x'(\tau)d\tau.$$

In contrast with the case when α is a constant, definitions of different types do not coincide.

Theorem 24 *The following relations between the left fractional operators hold:*

$$\substack{C \\ a}D_t^{\alpha(t)}x(t) = \substack{C \\ a}\mathbb{D}_t^{\alpha(t)}x(t) + \frac{\alpha'(t)}{\Gamma(2 - \alpha(t))}$$

$$\times \int_a^t (t - \tau)^{1-\alpha(t)}x'(\tau)\left[\frac{1}{1 - \alpha(t)} - \ln(t - \tau)\right]d\tau \quad (3.1)$$

and

$$\,^C_a D^{\alpha(t)}_t x(t) = \,^C_a \mathcal{D}^{\alpha(t)}_t x(t) - \frac{\alpha'(t)\Psi(1-\alpha(t))}{\Gamma(1-\alpha(t))}$$

$$\times \int_a^t (t-\tau)^{-\alpha(t)}[x(\tau) - x(a)]d\tau. \quad (3.2)$$

Proof Integrating by parts, one gets

$$\,^C_a D^{\alpha(t)}_t x(t) = \frac{1}{\Gamma(1-\alpha(t))} \frac{d}{dt} \int_a^t (t-\tau)^{-\alpha(t)}[x(\tau) - x(a)]d\tau$$

$$= \frac{1}{\Gamma(1-\alpha(t))} \frac{d}{dt} \left[\frac{1}{1-\alpha(t)} \int_a^t (t-\tau)^{1-\alpha(t)} x'(\tau)d\tau \right].$$

Differentiating the integral, it follows that

$$\,^C_a D^{\alpha(t)}_t x(t) = \frac{1}{\Gamma(1-\alpha(t))} \left[\frac{\alpha'(t)}{(1-\alpha(t))^2} \int_a^t (t-\tau)^{1-\alpha(t)} x'(\tau)d\tau \right.$$

$$+ \frac{1}{1-\alpha(t)} \int_a^t (t-\tau)^{1-\alpha(t)} x'(\tau) \left[-\alpha'(t)\ln(t-\tau) + \frac{1-\alpha(t)}{t-\tau} \right] d\tau \right]$$

$$= \,^C_a \mathcal{D}^{\alpha(t)}_t x(t) + \frac{\alpha'(t)}{\Gamma(2-\alpha(t))} \int_a^t (t-\tau)^{1-\alpha(t)} x'(\tau) \left[\frac{1}{1-\alpha(t)} - \ln(t-\tau) \right] d\tau.$$

The second formula follows from direct calculations.

Therefore, when the order $\alpha(t) \equiv c$ is a constant, or for constant functions $x(t) \equiv k$, we have

$$\,^C_a D^{\alpha(t)}_t x(t) = \,^C_a \mathcal{D}^{\alpha(t)}_t x(t) = \,^C_a \mathbb{D}^{\alpha(t)}_t x(t).$$

Similarly, we obtain the next result.

Theorem 25 *The following relations between the right fractional operators hold:*

$$\,^C_t D^{\alpha(t)}_b x(t) = \,^C_t \mathbb{D}^{\alpha(t)}_b x(t) + \frac{\alpha'(t)}{\Gamma(2-\alpha(t))}$$

$$\times \int_t^b (\tau-t)^{1-\alpha(t)} x'(\tau) \left[\frac{1}{1-\alpha(t)} - \ln(\tau-t) \right] d\tau$$

and

$$\,^C_t D^{\alpha(t)}_b x(t) = \,^C_t \mathcal{D}^{\alpha(t)}_b x(t) + \frac{\alpha'(t)\Psi(1-\alpha(t))}{\Gamma(1-\alpha(t))} \int_t^b (\tau-t)^{-\alpha(t)}[x(\tau) - x(b)]d\tau.$$

Theorem 26 *Let* $x \in C^1([a,b], \mathbb{R})$. *At* $t = a$

$$\,^C_a D^{\alpha(t)}_t x(t) = \,^C_a \mathcal{D}^{\alpha(t)}_t x(t) = \,^C_a \mathbb{D}^{\alpha(t)}_t x(t) = 0;$$

at $t = b$

$$_t^C D_b^{\alpha(t)} x(t) = _t^C \mathcal{D}_b^{\alpha(t)} x(t) = _t^C \mathbb{D}_b^{\alpha(t)} x(t) = 0.$$

Proof We start proving the third equality at the initial time $t = a$. We simply note that

$$\left| _a^C \mathbb{D}_t^{\alpha(t)} x(t) \right| \leq \frac{\|x'\|}{\Gamma(1 - \alpha(t))} \int_a^t (t - \tau)^{-\alpha(t)} d\tau = \frac{\|x'\|}{\Gamma(2 - \alpha(t))} (t - a)^{1-\alpha(t)},$$

which is zero at $t = a$. For the first equality at $t = a$, using Eq. (3.1), and the two next relations

$$\left| \int_a^t (t - \tau)^{1-\alpha(t)} \frac{x'(\tau)}{1 - \alpha(t)} d\tau \right| \leq \frac{\|x'\|}{(1 - \alpha(t))(2 - \alpha(t))} (t - a)^{2-\alpha(t)}$$

and

$$\left| \int_a^t (t - \tau)^{1-\alpha(t)} x'(\tau) \ln(t - \tau) d\tau \right|$$
$$\leq \frac{\|x'\|}{2 - \alpha(t)} (t - a)^{2-\alpha(t)} \left| \ln(t - a) - \frac{1}{2 - \alpha(t)} \right|,$$

this latter inequality obtained from integration by parts, we prove that $_a^C D_t^{\alpha(t)} x(t) = 0$ at $t = a$. Finally, we prove the second equality at $t = a$ by considering Eq. (3.2): performing an integration by parts, we get

$$\left| \int_a^t (t - \tau)^{-\alpha(t)} [x(\tau) - x(a)] d\tau \right| \leq \frac{\|x'\|}{(1 - \alpha(t))(2 - \alpha(t))} (t - a)^{2-\alpha(t)}$$

and so $_a^C \mathcal{D}_t^{\alpha(t)} x(t) = 0$ at $t = a$. The proof that the right fractional operators also vanish at the end point $t = b$ follows by similar arguments.

With some computations, a relationship between the Riemann–Liouville and the Caputo fractional derivatives is easily deduced:

$$_a D_t^{\alpha(t)} x(t) = _a^C D_t^{\alpha(t)} x(t) + \frac{x(a)}{\Gamma(1 - \alpha(t))} \frac{d}{dt} \int_a^t (t - \tau)^{-\alpha(t)} d\tau$$
$$= _a^C D_t^{\alpha(t)} x(t) + \frac{x(a)}{\Gamma(1 - \alpha(t))} (t - a)^{-\alpha(t)}$$
$$+ \frac{x(a)\alpha'(t)}{\Gamma(2 - \alpha(t))} (t - a)^{1-\alpha(t)} \left[\frac{1}{1 - \alpha(t)} - \ln(t - a) \right]$$

and

$$_a\mathcal{D}_t^{\alpha(t)}x(t) = {}_a^C\mathcal{D}_t^{\alpha(t)}x(t) + x(a)\frac{d}{dt}\left(\frac{1}{\Gamma(1-\alpha(t))}\int_a^t (t-\tau)^{-\alpha(t)}d\tau\right)$$

$$= {}_a^C\mathcal{D}_t^{\alpha(t)}x(t) + \frac{x(a)}{\Gamma(1-\alpha(t))}(t-a)^{-\alpha(t)}$$

$$+ \frac{x(a)\alpha'(t)}{\Gamma(2-\alpha(t))}(t-a)^{1-\alpha(t)}\left[\Psi(2-\alpha(t)) - \ln(t-a)\right].$$

For the right fractional operators, we have

$$_t D_b^{\alpha(t)}x(t) = {}_t^C D_b^{\alpha(t)}x(t) + \frac{x(b)}{\Gamma(1-\alpha(t))}(b-t)^{-\alpha(t)}$$

$$- \frac{x(b)\alpha'(t)}{\Gamma(2-\alpha(t))}(b-t)^{1-\alpha(t)}\left[\frac{1}{1-\alpha(t)} - \ln(b-t)\right]$$

and

$$_t\mathcal{D}_b^{\alpha(t)}x(t) = {}_t^C\mathcal{D}_b^{\alpha(t)}x(t) + \frac{x(b)}{\Gamma(1-\alpha(t))}(b-t)^{-\alpha(t)}$$

$$- \frac{x(b)\alpha'(t)}{\Gamma(2-\alpha(t))}(b-t)^{1-\alpha(t)}\left[\Psi(2-\alpha(t)) - \ln(b-t)\right].$$

Thus, it is immediate to conclude that if $x(a) = 0$, then

$$_aD_t^{\alpha(t)}x(t) = {}_a^C D_t^{\alpha(t)}x(t) \quad \text{and} \quad _a\mathcal{D}_t^{\alpha(t)}x(t) = {}_a^C\mathcal{D}_t^{\alpha(t)}x(t)$$

and if $x(b) = 0$, then

$$_tD_b^{\alpha(t)}x(t) = {}_t^C D_b^{\alpha(t)}x(t) \quad \text{and} \quad _t\mathcal{D}_b^{\alpha(t)}x(t) = {}_t^C\mathcal{D}_b^{\alpha(t)}x(t).$$

Next, we obtain formulas for the Caputo fractional derivatives of a power function.

Lemma 27 Let $x(t) = (t-a)^\gamma$ with $\gamma > 0$. Then,

$$_a^C D_t^{\alpha(t)}x(t) = \frac{\Gamma(\gamma+1)}{\Gamma(\gamma-\alpha(t)+1)}(t-a)^{\gamma-\alpha(t)}$$

$$- \alpha'(t)\frac{\Gamma(\gamma+1)}{\Gamma(\gamma-\alpha(t)+2)}(t-a)^{\gamma-\alpha(t)+1}$$

$$\times [\ln(t-a) - \Psi(\gamma-\alpha(t)+2) + \Psi(1-\alpha(t))],$$

$$_a^C\mathcal{D}_t^{\alpha(t)}x(t) = \frac{\Gamma(\gamma+1)}{\Gamma(\gamma-\alpha(t)+1)}(t-a)^{\gamma-\alpha(t)}$$

$$- \alpha'(t)\frac{\Gamma(\gamma+1)}{\Gamma(\gamma-\alpha(t)+2)}(t-a)^{\gamma-\alpha(t)+1}$$

$$\times [\ln(t-a) - \Psi(\gamma-\alpha(t)+2)],$$

$$\,_a^C\mathbb{D}_t^{\alpha(t)}x(t) = \frac{\Gamma(\gamma+1)}{\Gamma(\gamma-\alpha(t)+1)}(t-a)^{\gamma-\alpha(t)}.$$

Proof The formula for $\,_a^C D_t^{\alpha(t)}x(t)$ follows immediately from Samko and Ross [11]. For the second equality, one has

$$
\begin{aligned}
\,_a^C\mathcal{D}_t^{\alpha(t)}x(t) &= \frac{d}{dt}\left(\frac{1}{\Gamma(1-\alpha(t))}\int_a^t(t-\tau)^{-\alpha(t)}(\tau-a)^\gamma d\tau\right) \\
&= \frac{d}{dt}\left(\frac{1}{\Gamma(1-\alpha(t))}\int_a^t(t-a)^{-\alpha(t)}\left(1-\frac{\tau-a}{t-a}\right)^{-\alpha(t)}(\tau-a)^\gamma d\tau\right).
\end{aligned}
$$

With the change of variables $\tau - a = s(t - a)$, and with the help of the Beta function $B(\cdot, \cdot)$ (see Definition 3), we prove that

$$
\begin{aligned}
\,_a^C\mathcal{D}_t^{\alpha(t)}x(t) &= \frac{d}{dt}\left(\frac{(t-a)^{-\alpha(t)}}{\Gamma(1-\alpha(t))}\int_0^1(1-s)^{-\alpha(t)}s^\gamma(t-a)^{\gamma+1}ds\right) \\
&= \frac{d}{dt}\left(\frac{(t-a)^{\gamma-\alpha(t)+1}}{\Gamma(1-\alpha(t))}B(\gamma+1,1-\alpha(t))\right) \\
&= \frac{d}{dt}\left(\frac{\Gamma(\gamma+1)}{\Gamma(\gamma-\alpha(t)+2)}(t-a)^{\gamma-\alpha(t)+1}\right).
\end{aligned}
$$

We obtain the desired formula by differentiating this latter expression. The last equality follows in a similar way.

Analogous relations to those of Lemma 27, for the right Caputo fractional derivatives of variable-order, are easily obtained.

Lemma 28 Let $x(t) = (b-t)^\gamma$ with $\gamma > 0$. Then,

$$
\begin{aligned}
\,_t^C D_b^{\alpha(t)}x(t) &= \frac{\Gamma(\gamma+1)}{\Gamma(\gamma-\alpha(t)+1)}(b-t)^{\gamma-\alpha(t)} \\
&\quad + \alpha'(t)\frac{\Gamma(\gamma+1)}{\Gamma(\gamma-\alpha(t)+2)}(b-t)^{\gamma-\alpha(t)+1} \\
&\quad \times [\ln(b-t) - \Psi(\gamma-\alpha(t)+2) + \Psi(1-\alpha(t))], \\
\,_t^C\mathcal{D}_b^{\alpha(t)}x(t) &= \frac{\Gamma(\gamma+1)}{\Gamma(\gamma-\alpha(t)+1)}(b-t)^{\gamma-\alpha(t)} \\
&\quad + \alpha'(t)\frac{\Gamma(\gamma+1)}{\Gamma(\gamma-\alpha(t)+2)}(b-t)^{\gamma-\alpha(t)+1} \\
&\quad \times [\ln(b-t) - \Psi(\gamma-\alpha(t)+2)], \\
\,_t^C\mathbb{D}_b^{\alpha(t)}x(t) &= \frac{\Gamma(\gamma+1)}{\Gamma(\gamma-\alpha(t)+1)}(b-t)^{\gamma-\alpha(t)}.
\end{aligned}
$$

With Lemma 27 in mind, we immediately see that

$$\, _a^C D_t^{\alpha(t)} x(t) \neq \, _a^C \mathcal{D}_t^{\alpha(t)} x(t) \neq \, _a^C \mathbb{D}_t^{\alpha(t)} x(t).$$

Also, at least for the power function, it suggests that $_a^C \mathcal{D}_t^{\alpha(t)} x(t)$ may be a more suitable inverse operation of the fractional integral when the order is variable. For example, consider functions $x(t) = t^2$ and $y(t) = (1-t)^2$, and the fractional order $\alpha(t) = \frac{5t+1}{10}$, $t \in [0, 1]$. Then, $0.1 \leq \alpha(t) \leq 0.6$ for all t. Next we compare the fractional derivatives of x and y of order $\alpha(t)$ with the fractional derivatives of constant order $\alpha = 0.1$ and $\alpha = 0.6$. By Lemma 27, we know that the left Caputo fractional derivatives of order $\alpha(t)$ of x are given by

$$_0^C D_t^{\alpha(t)} x(t) = \frac{2}{\Gamma(3 - \alpha(t))} t^{2-\alpha(t)}$$
$$- \frac{t^{3-\alpha(t)}}{\Gamma(4 - \alpha(t))} \left[\ln(t) - \Psi(4 - \alpha(t)) + \Psi(1 - \alpha(t)) \right],$$

$$_0^C \mathcal{D}_t^{\alpha(t)} x(t) = \frac{2}{\Gamma(3 - \alpha(t))} t^{2-\alpha(t)} - \frac{t^{3-\alpha(t)}}{\Gamma(4 - \alpha(t))} \left[\ln(t) - \Psi(4 - \alpha(t)) \right],$$

$$_0^C \mathbb{D}_t^{\alpha(t)} x(t) = \frac{2}{\Gamma(3 - \alpha(t))} t^{2-\alpha(t)},$$

while by Lemma 28, the right Caputo fractional derivatives of order $\alpha(t)$ of y are given by

$$_t^C D_1^{\alpha(t)} y(t) = \frac{2(1 - t)^{2-\alpha(t)}}{\Gamma(3 - \alpha(t))}$$
$$+ \frac{(1 - t)^{3-\alpha(t)}}{\Gamma(4 - \alpha(t))} \left[\ln(1 - t) - \Psi(4 - \alpha(t)) + \Psi(1 - \alpha(t)) \right],$$

$$_t^C \mathcal{D}_1^{\alpha(t)} y(t) = \frac{2(1 - t)^{2-\alpha(t)}}{\Gamma(3 - \alpha(t))} + \frac{(1 - t)^{3-\alpha(t)}}{\Gamma(4 - \alpha(t))} \left[\ln(1 - t) - \Psi(4 - \alpha(t)) \right],$$

$$_t^C \mathbb{D}_1^{\alpha(t)} y(t) = \frac{2(1 - t)^{2-\alpha(t)}}{\Gamma(3 - \alpha(t))}.$$

For a constant order α, we have

$$_0^C D_t^{\alpha} x(t) = \frac{2}{\Gamma(3 - \alpha)} t^{2-\alpha} \quad \text{and} \quad _t^C D_1^{\alpha} y(t) = \frac{2}{\Gamma(3 - \alpha)} (1 - t)^{2-\alpha}.$$

The results can be seen in Fig. 3.1.

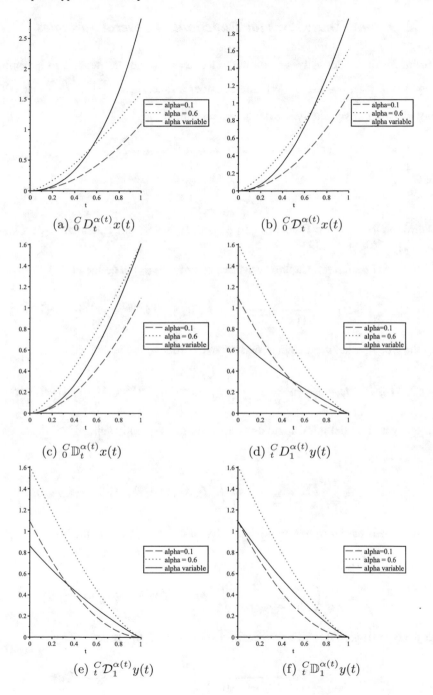

Fig. 3.1 Comparison between variable-order and constant-order fractional derivatives

3.1.2 Caputo Derivatives for Functions of Several Variables

Partial fractional derivatives are a natural extension and are defined in a similar way. Let $m \in \mathbb{N}$, $k \in \{1, \ldots, m\}$, and consider a function $x : \prod_{i=1}^{m}[a_i, b_i] \to \mathbb{R}$ with m variables. For simplicity, we define the vectors

$$[\tau]_k = (t_1, \ldots, t_{k-1}, \tau, t_{k+1}, \ldots, t_m) \in \mathbb{R}^m$$

and

$$(\bar{t}) = (t_1, \ldots, t_m) \in \mathbb{R}^m.$$

Definition 31 Given a function $x : \prod_{i=1}^{m}[a_i, b_i] \to \mathbb{R}$ and fractional orders $\alpha_k : [a_k, b_k] \to (0, 1)$, $k \in \{1, \ldots, m\}$,

1. the type I partial left Caputo derivative of order $\alpha_k(t_k)$ is defined by

$$_{a_k}^{C}D_{t_k}^{\alpha_k(t_k)} x(\bar{t}) = \frac{1}{\Gamma(1 - \alpha_k(t_k))} \frac{\partial}{\partial t_k} \int_{a_k}^{t_k} (t_k - \tau)^{-\alpha_k(t_k)} \left(x[\tau]_k - x[a_k]_k\right) d\tau;$$

2. the type I partial right Caputo derivative of order $\alpha_k(t_k)$ is defined by

$$_{t_k}^{C}D_{b_k}^{\alpha_k(t_k)} x(\bar{t}) = \frac{-1}{\Gamma(1 - \alpha_k(t_k))} \frac{\partial}{\partial t_k} \int_{t_k}^{b_k} (\tau - t_k)^{-\alpha_k(t_k)} \left(x[\tau]_k - x[b_k]_k\right) d\tau;$$

3. the type II partial left Caputo derivative of order $\alpha_k(t_k)$ is defined by

$$_{a_k}^{C}\mathcal{D}_{t_k}^{\alpha_k(t_k)} x(\bar{t})$$
$$= \frac{\partial}{\partial t_k} \left(\frac{1}{\Gamma(1 - \alpha_k(t_k))} \int_{a_k}^{t_k} (t_k - \tau)^{-\alpha_k(t_k)} \left(x[\tau]_k - x[a_k]_k\right) d\tau \right);$$

4. the type II partial right Caputo derivative of order $\alpha_k(t_k)$ is defined by

$$_{t_k}^{C}\mathcal{D}_{b_k}^{\alpha_k(t_k)} x(\bar{t})$$
$$= \frac{\partial}{\partial t_k} \left(\frac{-1}{\Gamma(1 - \alpha_k(t_k))} \int_{t_k}^{b_k} (\tau - t_k)^{-\alpha_k(t_k)} \left(x[\tau]_k - x[b_k]_k\right) d\tau \right);$$

5. the type III partial left Caputo derivative of order $\alpha_k(t_k)$ is defined by

$$_{a_k}^{C}\mathbb{D}_{t_k}^{\alpha_k(t_k)} x(\bar{t}) = \frac{1}{\Gamma(1 - \alpha_k(t_k))} \int_{a_k}^{t_k} (t_k - \tau)^{-\alpha_k(t_k)} \frac{\partial x}{\partial t_k}[\tau]_k d\tau;$$

6. the type III partial right Caputo derivative of order $\alpha_k(t_k)$ is defined by

$$\,^C_{t_k}\mathbb{D}^{\alpha_k(t_k)}_{b_k}x(\bar{t}) = \frac{-1}{\Gamma(1 - \alpha_k(t_k))} \int_{t_k}^{b_k} (\tau - t_k)^{-\alpha_k(t_k)} \frac{\partial x}{\partial t_k}[\tau]_k d\tau.$$

Similarly as done before, relations between these definitions can be proven.

Theorem 29 *The following four formulas hold:*

$$\,^C_{a_k} D^{\alpha_k(t_k)}_{t_k}x(\bar{t}) = \,^C_{a_k}\mathbb{D}^{\alpha_k(t_k)}_{t_k}x(\bar{t})$$
$$+ \frac{\alpha'_k(t_k)}{\Gamma(2 - \alpha_k(t_k))} \int_{a_k}^{t_k} (t_k - \tau)^{1-\alpha_k(t_k)} \frac{\partial x}{\partial t_k}[\tau]_k \left[\frac{1}{1 - \alpha_k(t_k)} - \ln(t_k - \tau)\right] d\tau, \quad (3.3)$$

$$\,^C_{a_k} D^{\alpha_k(t_k)}_{t_k}x(\bar{t}) = \,^C_{a_k}\mathcal{D}^{\alpha_k(t_k)}_{t_k}x(\bar{t})$$
$$- \frac{\alpha'_k(t_k)\Psi(1 - \alpha_k(t_k))}{\Gamma(1 - \alpha_k(t_k))} \int_{a_k}^{t_k} (t_k - \tau)^{-\alpha_k(t_k)}[x[\tau]_k - x[a_k]_k] d\tau, \quad (3.4)$$

$$\,^C_{t_k} D^{\alpha_k(t_k)}_{b_k}x(\bar{t}) = \,^C_{t_k}\mathbb{D}^{\alpha_k(t_k)}_{b_k}x(\bar{t})$$
$$+ \frac{\alpha'_k(t_k)}{\Gamma(2 - \alpha_k(t_k))} \int_{t_k}^{b_k} (\tau - t_k)^{1-\alpha_k(t_k)} \frac{\partial x}{\partial t_k}[\tau]_k \left[\frac{1}{1 - \alpha_k(t_k)} - \ln(\tau - t_k)\right] d\tau$$

and

$$\,^C_{t_k} D^{\alpha_k(t_k)}_{b_k}x(\bar{t}) = \,^C_{t_k}\mathcal{D}^{\alpha_k(t_k)}_{b_k}x(\bar{t})$$
$$+ \frac{\alpha'_k(t_k)\Psi(1 - \alpha_k(t_k))}{\Gamma(1 - \alpha_k(t_k))} \int_{t_k}^{b_k} (\tau - t_k)^{-\alpha_k(t_k)}[x[\tau]_k - x[b_k]_k] d\tau.$$

3.2 Numerical Approximations

Let $p \in \mathbb{N}$. We define

$$A_p = \frac{1}{\Gamma(p + 1 - \alpha_k(t_k))} \left[1 + \sum_{l=n-p+1}^{N} \frac{\Gamma(\alpha_k(t_k) - n + l)}{\Gamma(\alpha_k(t_k) - p)(l - n + p)!}\right],$$

$$B_p = \frac{\Gamma(\alpha_k(t_k) - n + p)}{\Gamma(1 - \alpha_k(t_k))\Gamma(\alpha_k(t_k))(p - n)!},$$

$$V_p(\bar{t}) = \int_{a_k}^{t_k} (\tau - a_k)^{p-n} \frac{\partial x}{\partial t_k}[\tau]_k d\tau,$$

$$L_p(\bar{t}) = \max_{\tau \in [a_k, t_k]} \left|\frac{\partial^p x}{\partial t_k^p}[\tau]_k\right|.$$

Theorem 30 *Let* $x \in C^{n+1}\left(\prod_{i=1}^{m}[a_i, b_i], \mathbb{R}\right)$ *with* $n \in \mathbb{N}$. *Then, for all* $k \in \{1, \ldots, m\}$ *and for all* $N \in \mathbb{N}$ *such that* $N \geq n$, *we have*

$$
{}_{a_k}^{C}\mathbb{D}_{t_k}^{\alpha_k(t_k)} x(\bar{t}) = \sum_{p=1}^{n} A_p (t_k - a_k)^{p-\alpha_k(t_k)} \frac{\partial^p x}{\partial t_k^p} [t_k]_k
$$

$$
+ \sum_{p=n}^{N} B_p (t_k - a_k)^{n-p-\alpha_k(t_k)} V_p(\bar{t}) + E(\bar{t}).
$$

The approximation error $E(\bar{t})$ *is bounded by*

$$
E(\bar{t}) \leq L_{n+1}(\bar{t}) \frac{\exp((n - \alpha_k(t_k))^2 + n - \alpha_k(t_k))}{\Gamma(n+1-\alpha_k(t_k))N^{n-\alpha_k(t_k)}(n - \alpha_k(t_k))} (t_k - a_k)^{n+1-\alpha_k(t_k)}.
$$

Proof By definition,

$$
{}_{a_k}^{C}\mathbb{D}_{t_k}^{\alpha_k(t_k)} x(\bar{t}) = \frac{1}{\Gamma(1 - \alpha_k(t_k))} \int_{a_k}^{t_k} (t_k - \tau)^{-\alpha_k(t_k)} \frac{\partial x}{\partial t_k} [\tau]_k d\tau
$$

and, integrating by parts with $u'(\tau) = (t_k - \tau)^{-\alpha_k(t_k)}$ and $v(\tau) = \frac{\partial x}{\partial t_k}[\tau]_k$, we deduce that

$$
{}_{a_k}^{C}\mathbb{D}_{t_k}^{\alpha_k(t_k)} x(\bar{t}) = \frac{(t_k - a_k)^{1-\alpha_k(t_k)}}{\Gamma(2 - \alpha_k(t_k))} \frac{\partial x}{\partial t_k} [a_k]_k
$$

$$
+ \frac{1}{\Gamma(2 - \alpha_k(t_k))} \int_{a_k}^{t_k} (t_k - \tau)^{1-\alpha_k(t_k)} \frac{\partial^2 x}{\partial t_k^2} [\tau]_k d\tau.
$$

Integrating again by parts, taking $u'(\tau) = (t_k - \tau)^{1-\alpha_k(t_k)}$ and $v(\tau) = \frac{\partial^2 x}{\partial t_k^2}[\tau]_k$, we get

$$
{}_{a_k}^{C}\mathbb{D}_{t_k}^{\alpha_k(t_k)} x(\bar{t}) = \frac{(t_k - a_k)^{1-\alpha_k(t_k)}}{\Gamma(2 - \alpha_k(t_k))} \frac{\partial x}{\partial t_k} [a_k]_k + \frac{(t_k - a_k)^{2-\alpha_k(t_k)}}{\Gamma(3 - \alpha_k(t_k))} \frac{\partial^2 x}{\partial t_k^2} [a_k]_k
$$

$$
+ \frac{1}{\Gamma(3 - \alpha_k(t_k))} \int_{a_k}^{t_k} (t_k - \tau)^{2-\alpha_k(t_k)} \frac{\partial^3 x}{\partial t_k^3} [\tau]_k d\tau.
$$

Repeating the same procedure $n - 2$ more times, we get the expansion formula

$$
{}_{a_k}^{C}\mathbb{D}_{t_k}^{\alpha_k(t_k)} x(\bar{t}) = \sum_{p=1}^{n} \frac{(t_k - a_k)^{p-\alpha_k(t_k)}}{\Gamma(p+1-\alpha_k(t_k))} \frac{\partial^p x}{\partial t_k^p} [a_k]_k
$$

$$
+ \frac{1}{\Gamma(n+1-\alpha_k(t_k))} \int_{a_k}^{t_k} (t_k - \tau)^{n-\alpha_k(t_k)} \frac{\partial^{n+1} x}{\partial t_k^{n+1}} [\tau]_k d\tau.
$$

Using the equalities

$$(t_k - \tau)^{n-\alpha_k(t_k)} = (t_k - a_k)^{n-\alpha_k(t_k)} \left(1 - \frac{\tau - a_k}{t_k - a_k} \right)^{n-\alpha_k(t_k)}$$

$$= (t_k - a_k)^{n-\alpha_k(t_k)} \left[\sum_{p=0}^{N} \binom{n-\alpha_k(t_k)}{p} (-1)^p \frac{(\tau - a_k)^p}{(t_k - a_k)^p} + \overline{E}(\bar{t}) \right]$$

with

$$\overline{E}(\bar{t}) = \sum_{p=N+1}^{\infty} \binom{n-\alpha_k(t_k)}{p} (-1)^p \frac{(\tau - a_k)^p}{(t_k - a_k)^p},$$

we arrive at

$$\begin{aligned}
{}_{a_k}^{C}\mathbb{D}_{t_k}^{\alpha_k(t_k)} x(\bar{t}) &= \sum_{p=1}^{n} \frac{(t_k - a_k)^{p-\alpha_k(t_k)}}{\Gamma(p+1-\alpha_k(t_k))} \frac{\partial^p x}{\partial t_k^p} [a_k]_k \\
&\quad + \frac{(t_k - a_k)^{n-\alpha_k(t_k)}}{\Gamma(n+1-\alpha_k(t_k))} \int_{a_k}^{t_k} \sum_{p=0}^{N} \binom{n-\alpha_k(t_k)}{p} (-1)^p \frac{(\tau - a_k)^p}{(t_k - a_k)^p} \frac{\partial^{n+1} x}{\partial t_k^{n+1}} [\tau]_k d\tau + E(\bar{t}) \\
&= \sum_{p=1}^{n} \frac{(t_k - a_k)^{p-\alpha_k(t_k)}}{\Gamma(p+1-\alpha_k(t_k))} \frac{\partial^p x}{\partial t_k^p} [a_k]_k + \frac{(t_k - a_k)^{n-\alpha_k(t_k)}}{\Gamma(n+1-\alpha_k(t_k))} \\
&\quad \times \sum_{p=0}^{N} \binom{n-\alpha_k(t_k)}{p} \frac{(-1)^p}{(t_k - a_k)^p} \int_{a_k}^{t_k} (\tau - a_k)^p \frac{\partial^{n+1} x}{\partial t_k^{n+1}} [\tau]_k d\tau + E(\bar{t})
\end{aligned}$$

with

$$E(\bar{t}) = \frac{(t_k - a_k)^{n-\alpha_k(t_k)}}{\Gamma(n+1-\alpha_k(t_k))} \int_{a_k}^{t_k} \overline{E}(\bar{t}) \frac{\partial^{n+1} x}{\partial t_k^{n+1}} [\tau]_k d\tau.$$

Now, we split the last sum into $p = 0$ and the remaining terms $p = 1, \ldots, N$ and integrate by parts with $u(\tau) = (\tau - a_k)^p$ and $v'(\tau) = \frac{\partial^{n+1} x}{\partial t_k^{n+1}} [\tau]_k$. Observing that

$$\binom{n-\alpha_k(t_k)}{p} (-1)^p = \frac{\Gamma(\alpha_k(t_k) - n + p)}{\Gamma(\alpha_k(t_k) - n) p!},$$

we obtain:

$$\frac{(t_k - a_k)^{n-\alpha_k(t_k)}}{\Gamma(n+1-\alpha_k(t_k))} \sum_{p=0}^{N} \binom{n-\alpha_k(t_k)}{p} \frac{(-1)^p}{(t_k - a_k)^p} \int_{a_k}^{t_k} (\tau - a_k)^p \frac{\partial^{n+1} x}{\partial t_k^{n+1}} [\tau]_k d\tau$$

$$= \frac{(t_k - a_k)^{n-\alpha_k(t_k)}}{\Gamma(n+1-\alpha_k(t_k))} \left[\frac{\partial^n x}{\partial t_k^n} [t_k]_k - \frac{\partial^n x}{\partial t_k^n} [a_k]_k \right]$$

$$+ \frac{(t_k - a_k)^{n - \alpha_k(t_k)}}{\Gamma(n + 1 - \alpha_k(t_k))} \sum_{p=1}^{N} \frac{\Gamma(\alpha_k(t_k) - n + p)}{\Gamma(\alpha_k(t_k) - n) p! (t_k - a_k)^p}$$

$$\times \left[(t_k - a_k)^p \frac{\partial^n x}{\partial t_k^n} [t_k]_k - \int_{a_k}^{t_k} p(\tau - a_k)^{p-1} \frac{\partial^n x}{\partial t_k^n} [\tau]_k d\tau \right]$$

$$= - \frac{(t_k - a_k)^{n - \alpha_k(t_k)}}{\Gamma(n + 1 - \alpha_k(t_k))} \frac{\partial^n x}{\partial t_k^n} [a_k]_k + \frac{(t_k - a_k)^{n - \alpha_k(t_k)}}{\Gamma(n + 1 - \alpha_k(t_k))} \frac{\partial^n x}{\partial t_k^n} [t_k]_k$$

$$\times \left[1 + \sum_{p=1}^{N} \frac{\Gamma(\alpha_k(t_k) - n + p)}{\Gamma(\alpha_k(t_k) - n) p!} \right] + \frac{(t_k - a_k)^{n - \alpha_k(t_k) - 1}}{\Gamma(n - \alpha_k(t_k))}$$

$$\times \sum_{p=1}^{N} \frac{\Gamma(\alpha_k(t_k) - n + p)}{\Gamma(\alpha_k(t_k) + 1 - n)(p - 1)!(t_k - a_k)^{p-1}} \int_{a_k}^{t_k} (\tau - a_k)^{p-1} \frac{\partial^n x}{\partial t_k^n} [\tau]_k d\tau.$$

Thus, we get

$$_{a_k}^{C} \mathbb{D}_{t_k}^{\alpha_k(t_k)} x(\bar{t}) = \sum_{p=1}^{n-1} \frac{(t_k - a_k)^{p - \alpha_k(t_k)}}{\Gamma(p + 1 - \alpha_k(t_k))} \frac{\partial^p x}{\partial t_k^p} [a_k]_k$$

$$+ \frac{(t_k - a_k)^{n - \alpha_k(t_k)}}{\Gamma(n + 1 - \alpha_k(t_k))} \frac{\partial^n x}{\partial t_k^n} [t_k]_k \left[1 + \sum_{p=1}^{N} \frac{\Gamma(\alpha_k(t_k) - n + p)}{\Gamma(\alpha_k(t_k) - n) p!} \right]$$

$$+ \frac{(t_k - a_k)^{n - \alpha_k(t_k) - 1}}{\Gamma(n - \alpha_k(t_k))} \sum_{p=1}^{N} \frac{\Gamma(\alpha_k(t_k) - n + p)}{\Gamma(\alpha_k(t_k) + 1 - n)(p - 1)!(t_k - a_k)^{p-1}}$$

$$\times \int_{a_k}^{t_k} (\tau - a_k)^{p-1} \frac{\partial^n x}{\partial t_k^n} [\tau]_k d\tau + E(\bar{t}).$$

Repeating the process $n - 1$ more times with respect to the last sum, that is, splitting the first term of the sum and integrating by parts the obtained result, we arrive to

$$_{a_k}^{C} \mathbb{D}_{t_k}^{\alpha_k(t_k)} x(\bar{t}) = \sum_{p=1}^{n} \frac{(t_k - a_k)^{p - \alpha_k(t_k)}}{\Gamma(p + 1 - \alpha_k(t_k))} \frac{\partial^p x}{\partial t_k^p} [t_k]_k$$

$$\times \left[1 + \sum_{l=n-p+1}^{N} \frac{\Gamma(\alpha_k(t_k) - n + l)}{\Gamma(\alpha_k(t_k) - p)(l - n + p)!} \right]$$

$$+ \sum_{p=n}^{N} \frac{\Gamma(\alpha_k(t_k) - n + p)}{\Gamma(1 - \alpha_k(t_k)) \Gamma(\alpha_k(t_k))(p - n)!} (t_k - a_k)^{n - p - \alpha_k(t_k)}$$

$$\times \int_{a_k}^{t_k} (\tau - a_k)^{p-n} \frac{\partial x}{\partial t_k} [\tau]_k d\tau + E(\bar{t}).$$

We now seek the upper-bound formula for $E(\bar{t})$. Using the two relations

$$\left| \frac{\tau - a_k}{t_k - a_k} \right| \leq 1, \ \ \text{if} \ \tau \in [a_k, t_k]$$

and

$$\left| \binom{n - \alpha_k(t_k)}{p} \right| \leq \frac{\exp((n - \alpha_k(t_k))^2 + n - \alpha_k(t_k))}{p^{n+1-\alpha_k(t_k)}},$$

we get

$$\overline{E}(\bar{t}) \leq \sum_{p=N+1}^{\infty} \frac{\exp((n - \alpha_k(t_k))^2 + n - \alpha_k(t_k))}{p^{n+1-\alpha_k(t_k)}}$$

$$\leq \int_N^{\infty} \frac{\exp((n - \alpha_k(t_k))^2 + n - \alpha_k(t_k))}{p^{n+1-\alpha_k(t_k)}} \, dp$$

$$= \frac{\exp((n - \alpha_k(t_k))^2 + n - \alpha_k(t_k))}{N^{n-\alpha_k(t_k)}(n - \alpha_k(t_k))}.$$

Then,

$$E(\bar{t}) \leq L_{n+1}(\bar{t}) \frac{\exp((n - \alpha_k(t_k))^2 + n - \alpha_k(t_k))}{\Gamma(n + 1 - \alpha_k(t_k)) N^{n-\alpha_k(t_k)}(n - \alpha_k(t_k))} (t_k - a_k)^{n+1-\alpha_k(t_k)}.$$

This concludes the proof.

Remark 31 In Theorem 30, we have

$$\lim_{N \to \infty} E(\bar{t}) = 0$$

for all $\bar{t} \in \prod_{i=1}^m [a_i, b_i]$ and $n \in \mathbb{N}$.

Theorem 32 *Let* $x \in C^{n+1}\left(\prod_{i=1}^m [a_i, b_i], \mathbb{R}\right)$ *with* $n \in \mathbb{N}$. *Then, for all* $k \in \{1, \ldots, m\}$ *and for all* $N \in \mathbb{N}$ *such that* $N \geq n$, *we have*

$$\underset{a_k}{^C} D_{t_k}^{\alpha_k(t_k)} x(\bar{t}) = \sum_{p=1}^n A_p (t_k - a_k)^{p-\alpha_k(t_k)} \frac{\partial^p x}{\partial t_k^p}[t_k]_k$$

$$+ \sum_{p=n}^N B_p (t_k - a_k)^{n-p-\alpha_k(t_k)} V_p(\bar{t}) + \frac{\alpha_k'(t_k)(t_k - a_k)^{1-\alpha_k(t_k)}}{\Gamma(2 - \alpha_k(t_k))}$$

$$\times \left[\left(\frac{1}{1 - \alpha_k(t_k)} - \ln(t_k - a_k) \right) \sum_{p=0}^N \binom{1-\alpha_k(t_k)}{p} \frac{(-1)^p}{(t_k - a_k)^p} V_{n+p}(\bar{t}) \right.$$

$$\left. + \sum_{p=0}^N \binom{1-\alpha_k(t_k)}{p} (-1)^p \sum_{r=1}^N \frac{1}{r(t_k - a_k)^{p+r}} V_{n+p+r}(\bar{t}) \right] + E(\bar{t}).$$

The approximation error $E(\bar{t})$ is bounded by

$$E(\bar{t}) \leq L_{n+1}(\bar{t}) \frac{\exp((n - \alpha_k(t_k))^2 + n - \alpha_k(t_k))}{\Gamma(n + 1 - \alpha_k(t_k)) N^{n-\alpha_k(t_k)} (n - \alpha_k(t_k))} (t_k - a_k)^{n+1-\alpha_k(t_k)}$$

$$+ \left| \alpha_k'(t_k) \right| L_1(\bar{t}) \frac{\exp((1 - \alpha_k(t_k))^2 + 1 - \alpha_k(t_k))}{\Gamma(2 - \alpha_k(t_k)) N^{1-\alpha_k(t_k)} (1 - \alpha_k(t_k))}$$

$$\times \left[\left| \frac{1}{1 - \alpha_k(t_k)} - \ln(t_k - a_k) \right| + \frac{1}{N} \right] (t_k - a_k)^{2-\alpha_k(t_k)}.$$

Proof Taking into account relation (3.3) and Theorem 30, we only need to expand the term

$$\frac{\alpha_k'(t_k)}{\Gamma(2 - \alpha_k(t_k))} \int_{a_k}^{t_k} (t_k - \tau)^{1-\alpha_k(t_k)} \frac{\partial x}{\partial t_k} [\tau]_k \left[\frac{1}{1 - \alpha_k(t_k)} - \ln(t_k - \tau) \right] d\tau. \quad (3.5)$$

Splitting the integral, and using the expansion formulas

$$(t_k - \tau)^{1-\alpha_k(t_k)} = (t_k - a_k)^{1-\alpha_k(t_k)} \left(1 - \frac{\tau - a_k}{t_k - a_k} \right)^{1-\alpha_k(t_k)}$$

$$= (t_k - a_k)^{1-\alpha_k(t_k)} \left[\sum_{p=0}^{N} \binom{1-\alpha_k(t_k)}{p} (-1)^p \frac{(\tau - a_k)^p}{(t_k - a_k)^p} + \overline{E}_1(\bar{t}) \right]$$

with

$$\overline{E}_1(\bar{t}) = \sum_{p=N+1}^{\infty} \binom{1-\alpha_k(t_k)}{p} (-1)^p \frac{(\tau - a_k)^p}{(t_k - a_k)^p}$$

and

$$\ln(t_k - \tau) = \ln(t_k - a_k) + \ln \left(1 - \frac{\tau - a_k}{t_k - a_k} \right)$$

$$= \ln(t_k - a_k) - \sum_{r=1}^{N} \frac{1}{r} \frac{(\tau - a_k)^r}{(t_k - a_k)^r} - \overline{E}_2(\bar{t})$$

with

$$\overline{E}_2(\bar{t}) = \sum_{r=N+1}^{\infty} \frac{1}{r} \frac{(\tau - a_k)^r}{(t_k - a_k)^r},$$

we conclude that (3.5) is equivalent to

$$\frac{\alpha_k'(t_k)}{\Gamma(2 - \alpha_k(t_k))} \left[\left(\frac{1}{1 - \alpha_k(t_k)} - \ln(t_k - a_k) \right) \int_{a_k}^{t_k} (t_k - \tau)^{1-\alpha_k(t_k)} \frac{\partial x}{\partial t_k} [\tau]_k d\tau \right.$$

$$
-\int_{a_k}^{t_k} (t_k - \tau)^{1-\alpha_k(t_k)} \ln\left(1 - \frac{\tau - a_k}{t_k - a_k}\right) \frac{\partial x}{\partial t_k}[\tau]_k d\tau \Bigg]
$$

$$
= \frac{\alpha_k'(t_k)}{\Gamma(2 - \alpha_k(t_k))} \left[\left(\frac{1}{1 - \alpha_k(t_k)} - \ln(t_k - a_k)\right)\right.
$$

$$
\times \int_{a_k}^{t_k} (t_k - a_k)^{1-\alpha_k(t_k)} \sum_{p=0}^{N} \binom{1-\alpha_k(t_k)}{p} (-1)^p \frac{(\tau - a_k)^p}{(t_k - a_k)^p} \frac{\partial x}{\partial t_k}[\tau]_k d\tau
$$

$$
+ \int_{a_k}^{t_k} (t_k - a_k)^{1-\alpha_k(t_k)} \sum_{p=0}^{N} \binom{1-\alpha_k(t_k)}{p} (-1)^p \frac{(\tau - a_k)^p}{(t_k - a_k)^p}
$$

$$
\times \sum_{r=1}^{N} \frac{1}{r} \frac{(\tau - a_k)^r}{(t_k - a_k)^r} \frac{\partial x}{\partial t_k}[\tau]_k d\tau \Bigg] + \frac{\alpha_k'(t_k)}{\Gamma(2 - \alpha_k(t_k))} \left[\left(\frac{1}{1 - \alpha_k(t_k)} - \ln(t_k - a_k)\right)\right.
$$

$$
\times \int_{a_k}^{t_k} (t_k - a_k)^{1-\alpha_k(t_k)} \overline{E}_1(\bar{t}) \frac{\partial x}{\partial t_k}[\tau]_k d\tau
$$

$$
+ \int_{a_k}^{t_k} (t_k - a_k)^{1-\alpha_k(t_k)} \overline{E}_1(\bar{t}) \overline{E}_2(\bar{t}) \frac{\partial x}{\partial t_k}[\tau]_k d\tau \Bigg]
$$

$$
= \frac{\alpha_k'(t_k)(t_k - a_k)^{1-\alpha_k(t_k)}}{\Gamma(2 - \alpha_k(t_k))} \left[\left(\frac{1}{1 - \alpha_k(t_k)} - \ln(t_k - a_k)\right) \sum_{p=0}^{N} \binom{1-\alpha_k(t_k)}{p}\right.
$$

$$
\times \frac{(-1)^p}{(t_k - a_k)^p} V_{n+p}(\bar{t}) + \sum_{p=0}^{N} \binom{1-\alpha_k(t_k)}{p} (-1)^p \sum_{r=1}^{N} \frac{1}{r(t_k - a_k)^{p+r}} V_{n+p+r}(\bar{t}) \Bigg]
$$

$$
+ \frac{\alpha_k'(t_k)(t_k - a_k)^{1-\alpha_k(t_k)}}{\Gamma(2 - \alpha_k(t_k))} \left[\left(\frac{1}{1 - \alpha_k(t_k)} - \ln(t_k - a_k)\right)\right.
$$

$$
\times \int_{a_k}^{t_k} \overline{E}_1(\bar{t}) \frac{\partial x}{\partial t_k}[\tau]_k d\tau + \int_{a_k}^{t_k} \overline{E}_1(\bar{t}) \overline{E}_2(\bar{t}) \frac{\partial x}{\partial t_k}[\tau]_k d\tau \Bigg].
$$

For the error analysis, we know from Theorem 30 that

$$
\overline{E}_1(\bar{t}) \leq \frac{\exp((1 - \alpha_k(t_k))^2 + 1 - \alpha_k(t_k))}{N^{1-\alpha_k(t_k)}(1 - \alpha_k(t_k))}.
$$

Then,

$$
\left|\int_{a_k}^{t_k} (t_k - a_k)^{1-\alpha_k(t_k)} \overline{E}_1(\bar{t}) \frac{\partial x}{\partial t_k}[\tau]_k d\tau\right|
$$

$$
\leq L_1(\bar{t}) \frac{\exp((1 - \alpha_k(t_k))^2 + 1 - \alpha_k(t_k))}{N^{1-\alpha_k(t_k)}(1 - \alpha_k(t_k))} (t_k - a_k)^{2-\alpha_k(t_k)}.
$$

(3.6)

On the other hand, we have

$$\left| \int_{a_k}^{t_k} (t_k - a_k)^{1-\alpha_k(t_k)} \overline{E}_1(\bar{t}) \overline{E}_2(\bar{t}) \frac{\partial x}{\partial t_k}[\tau]_k d\tau \right|$$

$$\leq L_1(\bar{t}) \frac{\exp((1-\alpha_k(t_k))^2 + 1 - \alpha_k(t_k))}{N^{1-\alpha_k(t_k)}(1-\alpha_k(t_k))} (t_k - a_k)^{1-\alpha_k(t_k)}$$

$$\times \sum_{r=N+1}^{\infty} \frac{1}{r(t_k - a_k)^r} \int_{a_k}^{t_k} (\tau - a_k)^r d\tau$$

$$= L_1(\bar{t}) \frac{\exp((1-\alpha_k(t_k))^2 + 1 - \alpha_k(t_k))}{N^{1-\alpha_k(t_k)}(1-\alpha_k(t_k))} (t_k - a_k)^{1-\alpha_k(t_k)} \sum_{r=N+1}^{\infty} \frac{t_k - a_k}{r(r+1)}$$

$$\leq L_1(\bar{t}) \frac{\exp((1-\alpha_k(t_k))^2 + 1 - \alpha_k(t_k))}{N^{2-\alpha_k(t_k)}(1-\alpha_k(t_k))} (t_k - a_k)^{2-\alpha_k(t_k)}. \tag{3.7}$$

We get the desired result by combining inequalities (3.6) and (3.7).

Theorem 33 *Let* $x \in C^{n+1}(\prod_{i=1}^{m}[a_i, b_i], \mathbb{R})$ *with* $n \in \mathbb{N}$. *Then, for all* $k \in \{1, \ldots, m\}$ *and for all* $N \in \mathbb{N}$ *such that* $N \geq n$, *we have*

$$_{a_k}^{C}\mathcal{D}_{t_k}^{\alpha_k(t_k)} x(\bar{t}) = \sum_{p=1}^{n} A_p (t_k - a_k)^{p-\alpha_k(t_k)} \frac{\partial^p x}{\partial t_k^p}[t_k]_k$$

$$+ \sum_{p=n}^{N} B_p (t_k - a_k)^{n-p-\alpha_k(t_k)} V_p(\bar{t}) + \frac{\alpha_k'(t_k)(t_k - a_k)^{1-\alpha_k(t_k)}}{\Gamma(2 - \alpha_k(t_k))}$$

$$\times \left[(\Psi(2 - \alpha_k(t_k)) - \ln(t_k - a_k)) \sum_{p=0}^{N} \binom{1-\alpha_k(t_k)}{p} \frac{(-1)^p}{(t_k - a_k)^p} V_{n+p}(\bar{t}) \right.$$

$$\left. + \sum_{p=0}^{N} \binom{1-\alpha_k(t_k)}{p} (-1)^p \sum_{r=1}^{N} \frac{1}{r(t_k - a_k)^{p+r}} V_{n+p+r}(\bar{t}) \right] + E(\bar{t}).$$

The approximation error $E(\bar{t})$ *is bounded by*

$$E(\bar{t}) \leq L_{n+1}(\bar{t}) \frac{\exp((n - \alpha_k(t_k))^2 + n - \alpha_k(t_k))}{\Gamma(n+1-\alpha_k(t_k))N^{n-\alpha_k(t_k)}(n - \alpha_k(t_k))} (t_k - a_k)^{n+1-\alpha_k(t_k)}$$

$$+ |\alpha_k'(t_k)| L_1(\bar{t}) \frac{\exp((1-\alpha_k(t_k))^2 + 1 - \alpha_k(t_k))}{\Gamma(2-\alpha_k(t_k))N^{1-\alpha_k(t_k)}(1 - \alpha_k(t_k))}$$

$$\times \left[|\Psi(2 - \alpha_k(t_k)) - \ln(t_k - a_k)| + \frac{1}{N} \right] (t_k - a_k)^{2-\alpha_k(t_k)}.$$

Proof Starting with relation (3.4), and integrating by parts the integral, we obtain that

$$\begin{aligned}
{}_{a_k}^{C}D_{t_k}^{\alpha_k(t_k)}x(\bar{t}) ={}& {}_{a_k}^{C}D_{t_k}^{\alpha_k(t_k)}x(\bar{t}) \\
&+ \frac{\alpha_k'(t_k)\Psi(1-\alpha_k(t_k))}{\Gamma(2-\alpha_k(t_k))}\int_{a_k}^{t_k}(t_k-\tau)^{1-\alpha_k(t_k)}\frac{\partial x}{\partial t_k}[\tau]_k d\tau.
\end{aligned}$$

The rest of the proof is similar to the one of Theorem 32.

Remark 34 As particular cases of Theorems 30, 32 and 33, we obtain expansion formulas for ${}_a^C D_t^{\alpha(t)}x(t)$, ${}_a^C\mathcal{D}_t^{\alpha(t)}x(t)$, and ${}_a^C\mathbb{D}_t^{\alpha(t)}x(t)$.

With respect to the three right fractional operators of Definition 31, we set, for $p \in \mathbb{N}$,

$$C_p = \frac{(-1)^p}{\Gamma(p+1-\alpha_k(t_k))}\left[1+\sum_{l=n-p+1}^{N}\frac{\Gamma(\alpha_k(t_k)-n+l)}{\Gamma(\alpha_k(t_k)-p)(l-n+p)!}\right],$$

$$D_p = \frac{-\Gamma(\alpha_k(t_k)-n+p)}{\Gamma(1-\alpha_k(t_k))\Gamma(\alpha_k(t_k))(p-n)!},$$

$$W_p(\bar{t}) = \int_{t_k}^{b_k}(b_k-\tau)^{p-n}\frac{\partial x}{\partial t_k}[\tau]_k d\tau,$$

$$M_p(\bar{t}) = \max_{\tau\in[t_k,b_k]}\left|\frac{\partial^p x}{\partial t_k^p}[\tau]_k\right|.$$

The expansion formulas are given in Theorems 35–37. We omit the proofs since they are similar to the corresponding left ones.

Theorem 35 *Let* $x \in C^{n+1}\left(\prod_{i=1}^{m}[a_i,b_i],\mathbb{R}\right)$ *with* $n \in \mathbb{N}$. *Then, for all* $k \in \{1,\ldots,m\}$ *and for all* $N \in \mathbb{N}$ *such that* $N \geq n$, *we have*

$$\begin{aligned}
{}_{t_k}^{C}\mathbb{D}_{b_k}^{\alpha_k(t_k)}x(\bar{t}) ={}& \sum_{p=1}^{n}C_p(b_k-t_k)^{p-\alpha_k(t_k)}\frac{\partial^p x}{\partial t_k^p}[t_k]_k \\
&+ \sum_{p=n}^{N}D_p(b_k-t_k)^{n-p-\alpha_k(t_k)}W_p(\bar{t})+E(\bar{t}).
\end{aligned}$$

The approximation error $E(\bar{t})$ *is bounded by*

$$E(\bar{t}) \leq M_{n+1}(\bar{t})\frac{\exp((n-\alpha_k(t_k))^2+n-\alpha_k(t_k))}{\Gamma(n+1-\alpha_k(t_k))N^{n-\alpha_k(t_k)}(n-\alpha_k(t_k))}(b_k-t_k)^{n+1-\alpha_k(t_k)}.$$

Theorem 36 *Let* $x \in C^{n+1}\left(\prod_{i=1}^{m}[a_i,b_i],\mathbb{R}\right)$ *with* $n \in \mathbb{N}$. *Then, for all* $k \in \{1,\ldots,m\}$ *and for all* $N \in \mathbb{N}$ *such that* $N \geq n$, *we have*

$$
{}^{C}_{t_k}D^{\alpha_k(t_k)}_{b_k}x(\bar{t}) = \sum_{p=1}^{n} C_p (b_k - t_k)^{p-\alpha_k(t_k)} \frac{\partial^p x}{\partial t_k^p}[t_k]_k
$$

$$
+ \sum_{p=n}^{N} D_p (b_k - t_k)^{n-p-\alpha_k(t_k)} W_p(\bar{t}) + \frac{\alpha_k'(t_k)(b_k - t_k)^{1-\alpha_k(t_k)}}{\Gamma(2 - \alpha_k(t_k))}
$$

$$
\times \left[\left(\frac{1}{1 - \alpha_k(t_k)} - \ln(b_k - t_k) \right) \sum_{p=0}^{N} \binom{1-\alpha_k(t_k)}{p} \frac{(-1)^p}{(b_k - t_k)^p} W_{n+p}(\bar{t}) \right.
$$

$$
\left. + \sum_{p=0}^{N} \binom{1-\alpha_k(t_k)}{p}(-1)^p \sum_{r=1}^{N} \frac{1}{r(b_k - t_k)^{p+r}} W_{n+p+r}(\bar{t}) \right] + E(\bar{t}).
$$

The approximation error $E(\bar{t})$ is bounded by

$$
E(\bar{t}) \leq M_{n+1}(\bar{t}) \frac{\exp((n - \alpha_k(t_k))^2 + n - \alpha_k(t_k))}{\Gamma(n + 1 - \alpha_k(t_k)) N^{n-\alpha_k(t_k)}(n - \alpha_k(t_k))}(b_k - t_k)^{n+1-\alpha_k(t_k)}
$$

$$
+ |\alpha_k'(t_k)| M_1(\bar{t}) \frac{\exp((1 - \alpha_k(t_k))^2 + 1 - \alpha_k(t_k))}{\Gamma(2 - \alpha_k(t_k)) N^{1-\alpha_k(t_k)}(1 - \alpha_k(t_k))}
$$

$$
\times \left[\left| \frac{1}{1 - \alpha_k(t_k)} - \ln(b_k - t_k) \right| + \frac{1}{N} \right] (b_k - t_k)^{2-\alpha_k(t_k)}.
$$

Theorem 37 *Let* $x \in C^{n+1}\left(\prod_{i=1}^{m}[a_i, b_i], \mathbb{R}\right)$ *with* $n \in \mathbb{N}$. *Then, for all* $k \in \{1, \ldots, m\}$ *and for all* $N \in \mathbb{N}$ *such that* $N \geq n$, *we have*

$$
{}^{C}_{t_k}D^{\alpha_k(t_k)}_{b_k}x(\bar{t}) = \sum_{p=1}^{n} C_p (b_k - t_k)^{p-\alpha_k(t_k)} \frac{\partial^p x}{\partial t_k^p}[t_k]_k
$$

$$
+ \sum_{p=n}^{N} D_p (b_k - t_k)^{n-p-\alpha_k(t_k)} W_p(\bar{t}) + \frac{\alpha_k'(t_k)(b_k - t_k)^{1-\alpha_k(t_k)}}{\Gamma(2 - \alpha_k(t_k))}
$$

$$
\times \left[(\Psi(2 - \alpha_k(t_k)) - \ln(b_k - t_k)) \sum_{p=0}^{N} \binom{1-\alpha_k(t_k)}{p} \frac{(-1)^p}{(b_k - t_k)^p} W_{n+p}(\bar{t}) \right.
$$

$$
\left. + \sum_{p=0}^{N} \binom{1-\alpha_k(t_k)}{p}(-1)^p \sum_{r=1}^{N} \frac{1}{r(b_k - t_k)^{p+r}} W_{n+p+r}(\bar{t}) \right] + E(\bar{t}).
$$

The approximation error $E(\bar{t})$ is bounded by

$$E(\bar{t}) \leq M_{n+1}(\bar{t}) \frac{\exp((n - \alpha_k(t_k))^2 + n - \alpha_k(t_k))}{\Gamma(n + 1 - \alpha_k(t_k))N^{n-\alpha_k(t_k)}(n - \alpha_k(t_k))}(b_k - t_k)^{n+1-\alpha_k(t_k)}$$

$$+ |\alpha_k'(t_k)| M_1(\bar{t}) \frac{\exp((1 - \alpha_k(t_k))^2 + 1 - \alpha_k(t_k))}{\Gamma(2 - \alpha_k(t_k))N^{1-\alpha_k(t_k)}(1 - \alpha_k(t_k))}$$

$$\times \left[|\Psi(2 - \alpha_k(t_k)) - \ln(b_k - t_k)| + \frac{1}{N} \right](b_k - t_k)^{2-\alpha_k(t_k)}.$$

3.3 Example

To test the accuracy of the proposed method, we compare the fractional derivative of a concrete given function with some numerical approximations of it. For $t \in [0, 1]$, let $x(t) = t^2$ be the test function. For the order of the fractional derivatives, we consider two cases:

$$\alpha(t) = \frac{50t + 49}{100} \quad \text{and} \quad \beta(t) = \frac{t + 5}{10}.$$

We consider the approximations given in Theorems 30–33, with a fixed $n = 1$ and $N \in \{2, 4, 6\}$. The error of approximating $f(t)$ by $\tilde{f}(t)$ is measured by $|f(t) - \tilde{f}(t)|$. See Figs. 3.2, 3.3, 3.4, 3.5, 3.6 and 3.7.

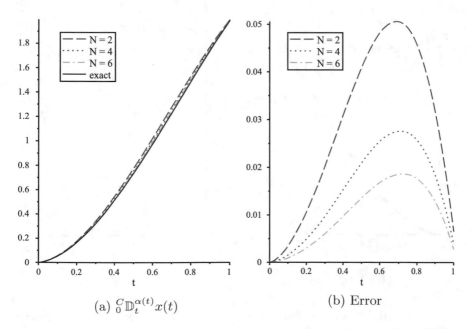

(a) $_{0}^{C}\mathbb{D}_t^{\alpha(t)} x(t)$

(b) Error

Fig. 3.2 Type III left Caputo derivative of order $\alpha(t)$ for the example of Sect. 3.3—analytic versus numerical approximations obtained from Theorem 30

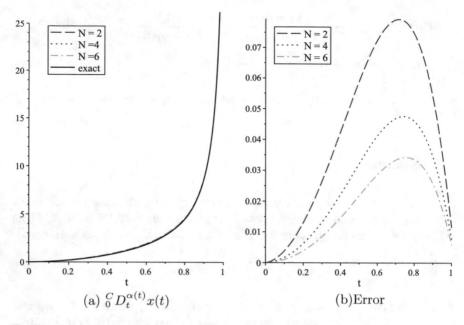

(a) $\,_0^C D_t^{\alpha(t)} x(t)$ (b)Error

Fig. 3.3 Type I left Caputo derivative of order $\alpha(t)$ for the example of Sect. 3.3—analytic versus numerical approximations obtained from Theorem 32

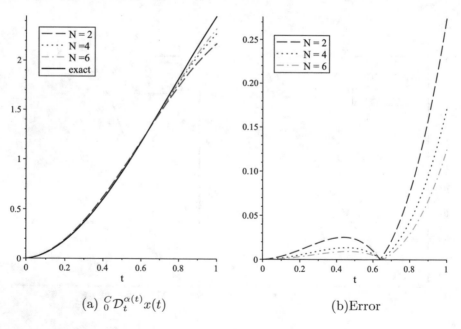

(a) $\,_0^C \mathcal{D}_t^{\alpha(t)} x(t)$ (b)Error

Fig. 3.4 Type II left Caputo derivative of order $\alpha(t)$ for the example of Sect. 3.3—analytic versus numerical approximations obtained from Theorem 33

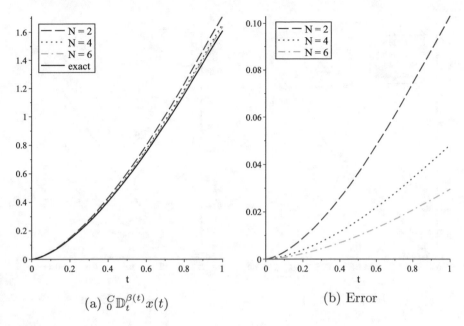

(a) $_0^C \mathbb{D}_t^{\beta(t)} x(t)$

(b) Error

Fig. 3.5 Type III left Caputo derivative of order $\beta(t)$ for the example of Sect. 3.3—analytic versus numerical approximations obtained from Theorem 30

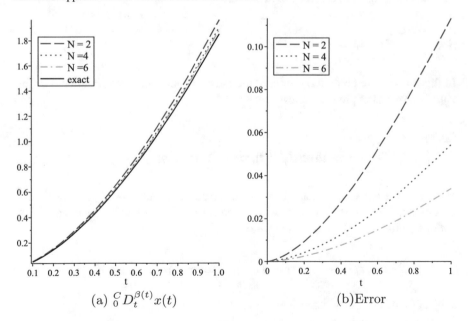

(a) $_0^C D_t^{\beta(t)} x(t)$

(b) Error

Fig. 3.6 Type I left Caputo derivative of order $\beta(t)$ for the example of Sect. 3.3—analytic versus numerical approximations obtained from Theorem 32

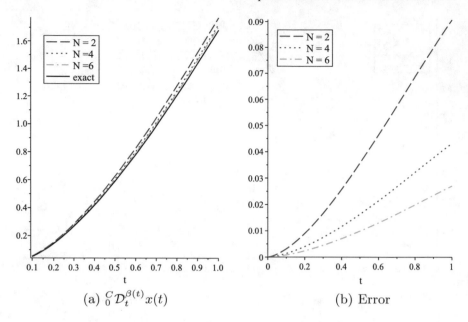

Fig. 3.7 Type II left Caputo derivative of order $\beta(t)$ for the example of Sect. 3.3—analytic versus numerical approximations obtained from Theorem 33

3.4   Applications

In this section, we apply the proposed technique to some concrete fractional differential equations of physical relevance.

3.4.1   A Time-Fractional Diffusion Equation

We extend the one-dimensional time-fractional diffusion equation [6] to the variable-order case. Consider $u = u(x, t)$ with domain $[0, 1]^2$. The partial fractional differential equation of order $\alpha(t)$ is the following:

$$\substack{C\\0}\mathbb{D}_t^{\alpha(t)} u(x, t) - \frac{\partial^2 u}{\partial x^2}(x, t) = f(x, t) \quad \text{for } x \in [0, 1], \quad t \in [0, 1], \tag{3.8}$$

subject to the boundary conditions

$$u(x, 0) = g(x), \quad \text{for } x \in (0, 1), \tag{3.9}$$

and

$$u(0, t) = u(1, t) = 0, \quad \text{for } t \in [0, 1]. \tag{3.10}$$

We mention that when $\alpha(t) \equiv 1$, one obtains the classical diffusion equation, and when $\alpha(t) \equiv 0$ one gets the classical Helmholtz elliptic equation. Using Lemma 27, it is easy to check that

$$u(x, t) = t^2 \sin(2\pi x)$$

is a solution to (3.8)–(3.10) with

$$f(x, t) = \left(\frac{2}{\Gamma(3 - \alpha(t))} t^{2 - \alpha(t)} + 4\pi^2 t^2 \right) \sin(2\pi x)$$

and

$$g(x) = 0$$

(compare with Example 1 in Lin and Xu [6]). The numerical procedure is the following: replace ${}_0^C\mathbb{D}_t^{\alpha(t)} u$ with the approximation given in Theorem 30, taking $n = 1$ and an arbitrary $N \geq 1$, that is,

$$
{}_0^C\mathbb{D}_t^{\alpha(t)} u(x, t) \approx A t^{1 - \alpha(t)} \frac{\partial u}{\partial t}(x, t) + \sum_{p=1}^{N} B_p t^{1 - p - \alpha(t)} V_p(x, t)
$$

with

$$
A = \frac{1}{\Gamma(2 - \alpha(t))} \left[1 + \sum_{l=1}^{N} \frac{\Gamma(\alpha(t) - 1 + l)}{\Gamma(\alpha(t) - 1)l!} \right],
$$

$$
B_p = \frac{\Gamma(\alpha(t) - 1 + p)}{\Gamma(1 - \alpha(t))\Gamma(\alpha(t))(p - 1)!},
$$

$$
V_p(x, t) = \int_0^t \tau^{p-1} \frac{\partial u}{\partial t}(x, \tau) d\tau.
$$

Then, the initial fractional problem (3.8)–(3.10) is approximated by the following system of second-order partial differential equations:

$$
A t^{1 - \alpha(t)} \frac{\partial u}{\partial t}(x, t) + \sum_{p=1}^{N} B_p t^{1 - p - \alpha(t)} V_p(x, t) - \frac{\partial^2 u}{\partial x^2}(x, t) = f(x, t)
$$

and

$$
\frac{\partial V_p}{\partial t}(x, t) = t^{p-1} \frac{\partial u}{\partial t}(x, t), \quad p = 1, \ldots, N,
$$

for $x \in [0, 1]$ and for $t \in [0, 1]$, subject to the boundary conditions

$$u(x, 0) = 0, \quad \text{for } x \in (0, 1),$$

$$u(0, t) = u(1, t) = 0, \quad \text{for } t \in [0, 1],$$

and

$$V_p(x, 0) = 0, \quad \text{for } x \in [0, 1], \quad p = 1, \dots, N.$$

3.4.2 A Fractional Partial Differential Equation in Fluid Mechanics

We now apply our approximation techniques to the following one-dimensional linear inhomogeneous fractional Burgers' equation of variable order (see Odibat and Momani [9, Example 5.2]):

$$
{}_0^C\mathbb{D}_t^{\alpha(t)} u(x, t) + \frac{\partial u}{\partial x}(x, t) - \frac{\partial^2 u}{\partial x^2}(x, t) = \frac{2t^{2-\alpha(t)}}{\Gamma(3 - \alpha(t))} + 2x - 2, \tag{3.11}
$$

for $x \in [0, 1]$ and $t \in [0, 1]$, subject to the boundary condition

$$u(x, 0) = x^2, \quad \text{for } x \in (0, 1). \tag{3.12}$$

Here,

$$F(x, t) = \frac{2t^{2-\alpha(t)}}{\Gamma(3 - \alpha(t))} + 2x - 2$$

is the external force field. Burgers' equation is used to model gas dynamics, traffic flow, turbulence, fluid mechanics, etc. The exact solution is

$$u(x, t) = x^2 + t^2.$$

The fractional problem (3.11)–(3.12) can be approximated by

$$
At^{1-\alpha(t)} \frac{\partial u}{\partial t}(x, t) + \sum_{p=1}^{N} B_p t^{1-p-\alpha(t)} V_p(x, t) + \frac{\partial u}{\partial x}(x, t) - \frac{\partial^2 u}{\partial x^2}(x, t)
$$

$$
= \frac{2t^{2-\alpha(t)}}{\Gamma(3 - \alpha(t))} + 2x - 2
$$

with A, B_p and V_p, $p \in \{1, \dots, N\}$, as in Sect. 3.4.1. The approximation error can be decreased as much as desired by increasing the value of N.

References

1. Caputo M (1967) Linear model of dissipation whose Q is almost frequency independent-II. Geophys J R Astr Soc 13:529–539
2. Chechkin AV, Gorenflo R, Sokolov IM (2005) Fractional diffusion in inhomogeneous media. J Phys A: Math Theor 38(42):L679–L684
3. Dalir M, Bashour M (2010) Applications of fractional calculus. Appl Math Sci 4(21–24):1021–1032
4. Diethelm K (2004) The analysis of fractional differential equations. Lecture notes in mathematics. Springer, Berlin
5. Džrbašjan MM, Nersesjan AB (1968) Fractional derivatives and the Cauchy problem for differential equations of fractional order. Izv Akad Nauk Armjan SSR Ser Math 3(1):3–29
6. Lin Y, Xu C (2007) Finite difference/spectral approximations for the time-fractional diffusion equation. J Comput Phys 225(2):1533–1552
7. Machado JAT, Silva MF, Barbosa RS, Jesus IS, Reis CM, Marcos MG, Galhano AF (2010) Some applications of fractional calculus in engineering. Math Probl Eng 2010. Art. ID 639801, 34 pp
8. Murio DA, Mejía CE (2008) Generalized time fractional IHCP with Caputo fractional derivatives. J Phys Conf Ser 135. Art. ID 012074, 8 pp
9. Odibat Z, Momani S (2009) The variational iteration method: an efficient scheme for handling fractional partial differential equations in fluid mechanics. Comput Math Appl 58(11–12):2199–2208
10. Rabotnov YN (1969) Creep problems in structural members. North-Holland series in applied mathematics and mechanics. Amsterdam, London
11. Samko SG, Ross B (1993) Integration and differentiation to a variable fractional order. Integr Transform Spec Funct 1(4):277–300
12. Samko SG, Kilbas AA, Marichev OI (1993) Fractional integrals and derivatives. Translated from the Russian original. Gordon and Breach, Yverdon (1987)
13. Santamaria F, Wils S, Schutter E, Augustine GJ (2006) Anomalous diffusion in Purkinje cell dendrites caused by spines. Neuron 52:635–648
14. Singh K, Saxena R, Kumar S (2013) Caputo-based fractional derivative in fractional Fourier transform domain. IEEE J Emerg Sel Top Circuits Syst 3:330–337
15. Sun HG, Chen W, Chen YQ (2009) Variable order fractional differential operators in anomalous diffusion modeling. Phys A 388(21):4586–4592
16. Sweilam NH, AL-Mrawm HM (2011) On the numerical solutions of the variable order fractional heat equation. Stud Nonlinear Sci 2:31–36
17. Tavares D, Almeida R, Torres DFM (2016) Caputo derivatives of fractional variable order: numerical approximations. Commun Nonlinear Sci Numer Simul 35:69–87
18. Yajima T, Yamasaki K (2012) Geometry of surfaces with Caputo fractional derivatives and applications to incompressible two-dimensional flows. J Phys A 45(6):065–201 15 pp

Chapter 4
The Fractional Calculus of Variations

In this chapter, we consider general fractional problems of the calculus of variations, where the Lagrangian depends on a combined Caputo fractional derivative of variable fractional order $^C D_\gamma^{\alpha(\cdot,\cdot),\beta(\cdot,\cdot)}$ given as a combination of the left and the right Caputo fractional derivatives of orders, respectively, $\alpha(\cdot,\cdot)$ and $\beta(\cdot,\cdot)$. More specifically, here we study some problems of the calculus of variations with integrands depending on the independent variable t, an arbitrary function x and a fractional derivative $^C D_\gamma^{\alpha(\cdot,\cdot),\beta(\cdot,\cdot)} x$ defined by

$$^C D_\gamma^{\alpha(\cdot,\cdot),\beta(\cdot,\cdot)} x(t) = \gamma_1 \, {}_a^C D_t^{\alpha(\cdot,\cdot)} x(t) + \gamma_2 \, {}_t^C D_b^{\beta(\cdot,\cdot)} x(t),$$

where $\gamma = (\gamma_1, \gamma_2) \in [0, 1]^2$, with γ_1 and γ_2 not both zero.

Starting from the fundamental variational problem, we investigate different types of variational problems: problems with time delay or with higher-order derivatives, isoperimetric problems, problems with holonomic constraints, and problems of Herglotz and those depending on combined Caputo fractional derivatives of variable order. Our variational problems are known as free-time problems because, in general, we impose a boundary condition at the initial time $t = a$, but we consider the endpoint b of the integral and the terminal state $x(b)$ to be free (variable). The main results provide necessary optimality conditions of Euler–Lagrange type, described by fractional differential equations of variable order and different transversality optimality conditions.

Our main contributions are to consider the Lagrangian containing a fractional operator where the order is not a constant, and may depend on time. Moreover, we do not only assume that $x(b)$ is free, but the endpoint b is also free.

In Sect. 4.1, we introduce the combined Caputo fractional derivative of variable order and provide the necessary concepts and results needed in the sequel

(Sect. 4.1.1). We deduce two formulas of integration by parts that involve the combined Caputo fractional derivative of higher order (Sect. 4.1.2).

The fractional variational problems under our consideration are formulated in terms of the fractional derivative $^C D_\gamma^{\alpha(\cdot,\cdot),\beta(\cdot,\cdot)}$. We discuss the fundamental concepts of a variational calculus such as the Euler–Lagrange equations and transversality conditions (Sect. 4.2), variational problems involving higher-order derivatives (Sect. 4.3), variational problems with time delay (Sect. 4.4), isoperimetric problems (Sect. 4.5), problems with holonomic constraints (Sect. 4.6), and the last Sect. 4.7 investigates fractional variational Herglotz problems.

Some illustrative examples are presented for all considered variational problems. The results of this chapter first appeared in [20–23].

4.1   Introduction

In this section, we recall the fundamental definition of the combined Caputo fractional derivative presented in Sect. 1.3.3 (see Definition 15) and generalize it for fractional variable order. In the end, we prove an integration by parts formula, involving the higher-order Caputo fractional derivative of variable order.

4.1.1   Combined Operators of Variable Order

Motivated by the combined fractional Caputo derivative, we propose the following definitions about combined variable-order fractional calculus.

Let $\alpha, \beta : [a, b]^2 \to (0, 1)$ be two functions and $\gamma = (\gamma_1, \gamma_2) \in [0, 1]^2$ a vector, with γ_1 and γ_2 not both zero.

Definition 32 The combined Riemann–Liouville fractional derivative of variable order, denoted by $D_\gamma^{\alpha(\cdot,\cdot),\beta(\cdot,\cdot)}$, is defined by

$$D_\gamma^{\alpha(\cdot,\cdot),\beta(\cdot,\cdot)} = \gamma_1 \, {}_a D_t^{\alpha(\cdot,\cdot)} + \gamma_2 \, {}_t D_b^{\beta(\cdot,\cdot)},$$

acting on $x \in C([a, b]; \mathbb{R})$ in the following way:

$$D_\gamma^{\alpha(\cdot,\cdot),\beta(\cdot,\cdot)} x(t) = \gamma_1 \, {}_a D_t^{\alpha(\cdot,\cdot)} x(t) + \gamma_2 \, {}_t D_b^{\beta(\cdot,\cdot)} x(t).$$

Definition 33 The combined Caputo fractional derivative operator of variable order, denoted by $^C D_\gamma^{\alpha(\cdot,\cdot),\beta(\cdot,\cdot)}$, is defined by

$$^C D_\gamma^{\alpha(\cdot,\cdot),\beta(\cdot,\cdot)} = \gamma_1 \, {}_a^C D_t^{\alpha(\cdot,\cdot)} + \gamma_2 \, {}_t^C D_b^{\beta(\cdot,\cdot)},$$

acting on $x \in C^1([a, b]; \mathbb{R})$ in the following way:

$$^C D_\gamma^{\alpha(\cdot,\cdot),\beta(\cdot,\cdot)} x(t) = \gamma_1\, {}^C_a D_t^{\alpha(\cdot,\cdot)} x(t) + \gamma_2\, {}^C_t D_b^{\beta(\cdot,\cdot)} x(t).$$

In the sequel, we need the auxiliary notation of the dual fractional derivative, defined by

$$D_{\overline{\gamma}}^{\beta(\cdot,\cdot),\alpha(\cdot,\cdot)} = \gamma_2\, {}_a D_t^{\beta(\cdot,\cdot)} + \gamma_1\, {}_t D_T^{\alpha(\cdot,\cdot)}, \tag{4.1}$$

where $\overline{\gamma} = (\gamma_2, \gamma_1)$ and $T \in (a, b]$.

To generalize the concept of the combined fractional derivative to higher orders, we need to review some definitions of higher-order operators.

Let $n \in \mathbb{N}$ and $x : [a, b] \to \mathbb{R}$ be a function of class C^n. The fractional order is a continuous function of two variables, $\alpha_n : [a, b]^2 \to (n - 1, n)$.

Definition 34 The left and right Riemann–Liouville fractional integrals of order $\alpha_n(\cdot, \cdot)$ are defined, respectively, by

$$_a I_t^{\alpha_n(\cdot,\cdot)} x(t) = \int_a^t \frac{1}{\Gamma(\alpha_n(t, \tau))} (t - \tau)^{\alpha_n(t,\tau)-1} x(\tau) d\tau$$

and

$$_t I_b^{\alpha_n(\cdot,\cdot)} x(t) = \int_t^b \frac{1}{\Gamma(\alpha_n(\tau, t))} (\tau - t)^{\alpha_n(\tau,t)-1} x(\tau) d\tau.$$

With this definition of Riemann–Liouville fractional integrals of variable order, and considering $n = 1$, in Appendix A.1 we implemented two functions leftFI(x,alpha,a) and rightFI(x,alpha,b) that approximate, respectively, the Riemann–Liouville fractional integrals $_a I_t^{\alpha_n(\cdot,\cdot)} x$ and $_t I_b^{\alpha_n(\cdot,\cdot)} x$, using the open-source software package **Chebfun** [24]. With these two functions, we present also, in Appendix A.1, an illustrative example where we determine computational approximations to the fractional integrals for a specific power function of the form $x(t) = t^\gamma$ (see Example 4.4 in Appendix A.1). For this numerical computation, we have used **MATLAB** [14].

Definition 35 The left and right Riemann–Liouville fractional derivatives of order $\alpha_n(\cdot, \cdot)$ are defined by

$$_a D_t^{\alpha_n(\cdot,\cdot)} x(t) = \frac{d^n}{dt^n} \int_a^t \frac{1}{\Gamma(n - \alpha_n(t, \tau))} (t - \tau)^{n-1-\alpha_n(t,\tau)} x(\tau) d\tau$$

and

$$_t D_b^{\alpha_n(\cdot,\cdot)} x(t) = (-1)^n \frac{d^n}{dt^n} \int_t^b \frac{1}{\Gamma(n - \alpha_n(\tau, t))} (\tau - t)^{n-1-\alpha_n(\tau,t)} x(\tau) d\tau,$$

respectively.

Definition 36 The left and right Caputo fractional derivatives of order $\alpha_n(\cdot, \cdot)$ are defined by

$$\,_a^C D_t^{\alpha_n(\cdot,\cdot)} x(t) = \int_a^t \frac{1}{\Gamma(n - \alpha_n(t, \tau))} (t - \tau)^{n-1-\alpha_n(t,\tau)} x^{(n)}(\tau) d\tau \qquad (4.2)$$

and

$$\,_t^C D_b^{\alpha_n(\cdot,\cdot)} x(t) = (-1)^n \int_t^b \frac{1}{\Gamma(n - \alpha_n(\tau, t))} (\tau - t)^{n-1-\alpha_n(\tau,t)} x^{(n)}(\tau) d\tau, \qquad (4.3)$$

respectively.

Remark 38 Definitions 35 and 36, for the particular case of order between 0 and 1, can be found in Malinowska et al. [16]. They seem to be new for the higher-order case.

Considering Definition 36, in Appendix A.2 we implement two new functions `leftCaputo(x,alpha,a,n)` and `rightCaputo(x,alpha,b,n)` that approximate the higher-order Caputo fractional derivatives $\,_a^C D_t^{\alpha_n(\cdot,\cdot)} x$ and $\,_t^C D_b^{\alpha_n(\cdot,\cdot)} x$, respectively. With these two functions, we present, also in Appendix A.2, an illustrative example where we study approximations to the Caputo fractional derivatives for a specific power function of the form $x(t) = t^\gamma$ (see Example 4.5 in Appendix A.2).

Remark 39 From Definition 34, it follows that

$$\,_a D_t^{\alpha_n(\cdot,\cdot)} x(t) = \frac{d^n}{dt^n} \,_a I_t^{n-\alpha_n(\cdot,\cdot)} x(t), \quad \,_t D_b^{\alpha_n(\cdot,\cdot)} x(t) = (-1)^n \frac{d^n}{dt^n} \,_t I_b^{n-\alpha_n(\cdot,\cdot)} x(t)$$

and

$$\,_a^C D_t^{\alpha_n(\cdot,\cdot)} x(t) = \,_a I_t^{n-\alpha_n(\cdot,\cdot)} \frac{d^n}{dt^n} x(t), \quad \,_t^C D_b^{\alpha_n(\cdot,\cdot)} x(t) = (-1)^n \,_t I_b^{n-\alpha_n(\cdot,\cdot)} \frac{d^n}{dt^n} x(t).$$

In Lemma 40, we obtain higher-order Caputo fractional derivatives of a power function. For that, we assume that the fractional order depends only on the first variable: $\alpha_n(t, \tau) := \overline{\alpha}_n(t)$, where $\overline{\alpha}_n : [a, b] \rightarrow (n - 1, n)$ is a given function.

Lemma 40 *Let* $x(t) = (t - a)^\gamma$ *with* $\gamma > n - 1$. *Then,*

$$\,_a^C D_t^{\overline{\alpha}_n(t)} x(t) = \frac{\Gamma(\gamma + 1)}{\Gamma(\gamma - \overline{\alpha}_n(t) + 1)} (t - a)^{\gamma - \overline{\alpha}_n(t)}.$$

Proof As $x(t) = (t - a)^\gamma$, if we differentiate it n times, we obtain

$$x^{(n)}(t) = \frac{\Gamma(\gamma + 1)}{\Gamma(\gamma - n + 1)} (t - a)^{\gamma - n}.$$

Using Definition 36 of the left Caputo fractional derivative, we get

$$
\begin{aligned}
{}_a^C D_t^{\overline{\alpha}_n(t)} x(t) &= \int_a^t \frac{1}{\Gamma(n - \overline{\alpha}_n(t))} (t - \tau)^{n-1-\overline{\alpha}_n(t)} x^{(n)}(\tau) d\tau \\
&= \int_a^t \frac{\Gamma(\gamma + 1)}{\Gamma(\gamma - n + 1)\Gamma(n - \overline{\alpha}_n(t))} (t - \tau)^{n-1-\overline{\alpha}_n(t)} (\tau - a)^{\gamma - n} d\tau.
\end{aligned}
$$

Now, we proceed with the change of variables $\tau - a = s(t - a)$. Using the Beta function $B(\cdot, \cdot)$, we obtain that

$$
\begin{aligned}
{}_a^C D_t^{\overline{\alpha}_n(t)} x(t) &= \frac{\Gamma(\gamma + 1)}{\Gamma(\gamma - n + 1)\Gamma(n - \overline{\alpha}_n(t))} \\
&\quad \times \int_0^1 (1 - s)^{n-1-\overline{\alpha}_n(t)} s^{\gamma - n} (t - a)^{\gamma - \overline{\alpha}_n(t)} ds \\
&= \frac{\Gamma(\gamma + 1)(t - a)^{\gamma - \overline{\alpha}_n(t)}}{\Gamma(\gamma - n + 1)\Gamma(n - \overline{\alpha}_n(t))} B(\gamma - n + 1, n - \overline{\alpha}_n(t)) \\
&= \frac{\Gamma(\gamma + 1)}{\Gamma(\gamma - \overline{\alpha}_n(t) + 1)} (t - a)^{\gamma - \overline{\alpha}_n(t)}.
\end{aligned}
$$

The proof is complete.

Considering the higher-order left Caputo fractional derivative's formula of a power function of the form $x(t) = (t - a)^\gamma$, deduced before, in Appendix A.2 we determine the left Caputo fractional derivative of the particular function $x(t) = t^4$ for several values of t and compare them with the approximated values obtained by the Chebfun function leftCaputo(x,alpha,a,n) (see Example 4.7 in Appendix A.2).

For our next result, we assume that the fractional order depends only on the second variable: $\alpha_n(\tau, t) := \overline{\alpha}_n(t)$, where $\overline{\alpha}_n : [a, b] \to (n - 1, n)$ is a given function. The proof is similar to that of Lemma 40, and so we omit it here.

Lemma 41 Let $x(t) = (b - t)^\gamma$ with $\gamma > n - 1$. Then,

$$
{}_t^C D_b^{\overline{\alpha}_n(t)} x(t) = \frac{\Gamma(\gamma + 1)}{\Gamma(\gamma - \overline{\alpha}_n(t) + 1)} (b - t)^{\gamma - \overline{\alpha}_n(t)}.
$$

The next step is to consider a linear combination of the previous fractional derivatives to define the combined fractional operators for higher-order.

Let $\alpha_n, \beta_n : [a, b]^2 \to (n - 1, n)$ be two variable fractional orders, $\gamma^n = (\gamma_1^n, \gamma_2^n) \in [0, 1]^2$ a vector, with γ_1 and γ_2 not both zero, and $x \in C^n([a, b]; \mathbb{R})$ a function.

Definition 37 The higher-order combined Riemann–Liouville fractional derivative is defined by

$$
D_{\gamma^n}^{\alpha_n(\cdot,\cdot),\beta_n(\cdot,\cdot)} = \gamma_1^n {}_a D_t^{\alpha_n(\cdot,\cdot)} + \gamma_2^n {}_t D_b^{\beta_n(\cdot,\cdot)},
$$

acting on $x \in C^n([a, b]; \mathbb{R})$ in the following way:

$$D_{\gamma^n}^{\alpha_n(\cdot,\cdot),\beta_n(\cdot,\cdot)}x(t) = \gamma_1^n {}_aD_t^{\alpha_n(\cdot,\cdot)}x(t) + \gamma_2^n {}_tD_b^{\beta_n(\cdot,\cdot)}x(t).$$

In our work, we use both Riemann–Liouville and Caputo derivatives definitions. The emphasis is, however, in Caputo fractional derivatives.

Definition 38 The higher-order combined Caputo fractional derivative of x at t is defined by

$$^C D_{\gamma^n}^{\alpha_n(\cdot,\cdot),\beta_n(\cdot,\cdot)}x(t) = \gamma_1^n {}_a^C D_t^{\alpha_n(\cdot,\cdot)}x(t) + \gamma_2^n {}_t^C D_b^{\beta_n(\cdot,\cdot)}x(t).$$

Similarly, in the sequel of this work, we need the auxiliary notation of the dual fractional derivative:

$$D_{\overline{\gamma^i}}^{\beta_i(\cdot,\cdot),\alpha_i(\cdot,\cdot)} = \gamma_2^i {}_aD_t^{\beta_i(\cdot,\cdot)} + \gamma_1^i {}_tD_T^{\alpha_i(\cdot,\cdot)}, \tag{4.4}$$

where $\overline{\gamma^i} = (\gamma_2^i, \gamma_1^i)$ and $T \in (a, b]$.

Some computational aspects about the combined Caputo fractional derivative of variable order are also discussed in Appendix A.3, using the software package **Chebfun**. For that, we developed the new function (see Example 4.8 in Appendix A.3) `combinedCaputo(x,alpha,beta,gamma1,gamma2,a,b,n)` and obtained approximated values for a particular power function.

4.1.2   Generalized Fractional Integration by Parts

When dealing with variational problems, one key property is integration by parts. Formulas of integration by parts have an important role in the proof of Euler–Lagrange conditions. In the following theorem, such formulas are proved for integrals involving higher-order Caputo fractional derivatives of variable order.

Let $n \in \mathbb{N}$ and $x, y \in C^n([a, b]; \mathbb{R})$ be two functions. The fractional order is a continuous function of two variables, $\alpha_n : [a, b]^2 \to (n - 1, n)$.

Theorem 42 *The higher-order Caputo fractional derivatives of variable order satisfy the integration by parts formulas*

$$\int_a^b y(t) {}_a^C D_t^{\alpha_n(\cdot,\cdot)}x(t)dt = \int_a^b x(t) {}_tD_b^{\alpha_n(\cdot,\cdot)}y(t)dt$$

$$+ \left[\sum_{k=0}^{n-1}(-1)^k x^{(n-1-k)}(t) \frac{d^k}{dt^k} {}_tI_b^{n-\alpha_n(\cdot,\cdot)}y(t) \right]_{t=a}^{t=b}$$

and

$$\int_a^b y(t) \, {}_t^C D_b^{\alpha_n(\cdot,\cdot)} x(t) dt = \int_a^b x(t) \, {}_a D_t^{\alpha_n(\cdot,\cdot)} y(t) dt$$

$$+ \left[\sum_{k=0}^{n-1} (-1)^{n+k} x^{(n-1-k)}(t) \frac{d^k}{dt^k} \, {}_a I_t^{n-\alpha_n(\cdot,\cdot)} y(t) \right]_{t=a}^{t=b}.$$

Proof Considering the Definition 36 of left Caputo fractional derivatives of order $\alpha_n(\cdot, \cdot)$, we obtain

$$\int_a^b y(t) \, {}_a^C D_t^{\alpha_n(\cdot,\cdot)} x(t) dt$$

$$= \int_a^b \int_a^t y(t) \frac{1}{\Gamma(n - \alpha_n(t, \tau))} (t - \tau)^{n-1-\alpha_n(t,\tau)} x^{(n)}(\tau) d\tau dt.$$

Using Dirichlet's formula, we rewrite it as

$$\int_a^b \int_t^b y(\tau) \frac{(\tau - t)^{n-1-\alpha_n(\tau,t)}}{\Gamma(n - \alpha_n(\tau, t))} x^{(n)}(t) d\tau dt$$

$$= \int_a^b x^{(n)}(t) \int_t^b \frac{(\tau - t)^{n-1-\alpha_n(\tau,t)}}{\Gamma(n - \alpha_n(\tau, t))} y(\tau) d\tau dt = \int_a^b x^{(n)}(t) \, {}_t I_b^{n-\alpha_n(\cdot,\cdot)} y(t) dt.$$

$$(4.5)$$

Using the (usual) integrating by parts formula, we get that (4.5) is equal to

$$- \int_a^b x^{(n-1)}(t) \frac{d}{dt} \, {}_t I_b^{n-\alpha_n(\cdot,\cdot)} y(t) dt + \left[x^{(n-1)}(t) \, {}_t I_b^{n-\alpha_n(\cdot,\cdot)} y(t) \right]_{t=a}^{t=b}.$$

Integrating by parts again, we obtain

$$\int_a^b x^{(n-2)}(t) \frac{d^2}{dt^2} \, {}_t I_b^{n-\alpha_n(\cdot,\cdot)} y(t) dt$$

$$+ \left[x^{(n-1)}(t) \, {}_t I_b^{n-\alpha_n(\cdot,\cdot)} y(t) - x^{(n-2)}(t) \frac{d}{dt} \, {}_t I_b^{n-\alpha_n(\cdot,\cdot)} y(t) \right]_{t=a}^{t=b}.$$

If we repeat this process $n - 2$ times more, we get

$$\int_a^b x(t)(-1)^n \frac{d^n}{dt^n} \, {}_t I_b^{n-\alpha_n(\cdot,\cdot)} y(t) dt$$

$$+ \left[\sum_{k=0}^{n-1} (-1)^k x^{(n-1-k)}(t) \frac{d^k}{dt^k} \, {}_t I_b^{n-\alpha_n(\cdot,\cdot)} y(t) \right]_{t=a}^{t=b}$$

$$= \int_a^b x(t) \, {}_t D_b^{\alpha_n(\cdot,\cdot)} y(t) dt + \left[\sum_{k=0}^{n-1} (-1)^k x^{(n-1-k)}(t) \frac{d^k}{dt^k} \, {}_t I_b^{n-\alpha_n(\cdot,\cdot)} y(t) \right]_{t=a}^{t=b}.$$

The second relation of the theorem for the right Caputo fractional derivative of order $\alpha_n(\cdot, \cdot)$ follows directly from the first by Caputo–Torres duality [4].

Remark 43 If we consider in Theorem 42 the particular case when $n = 1$, then the fractional integration by parts formulas takes the well-known forms presented in Theorem 13.

Remark 44 If x is such that $x^{(i)}(a) = x^{(i)}(b) = 0, i = 0, \ldots, n - 1$, then the higher-order formulas of fractional integration by parts given by Theorem 42 can be rewritten as

$$\int_a^b y(t) \, {}^C_a D_t^{\alpha_n(\cdot,\cdot)} x(t) dt = \int_a^b x(t) \, {}_t D_b^{\alpha_n(\cdot,\cdot)} y(t) dt$$

and

$$\int_a^b y(t) \, {}^C_t D_b^{\alpha_n(\cdot,\cdot)} x(t) dt = \int_a^b x(t) \, {}_a D_t^{\alpha_n(\cdot,\cdot)} y(t) dt.$$

4.2 Fundamental Variational Problem

This section is dedicated to establish necessary optimality conditions for variational problems with a Lagrangian depending on a combined Caputo derivative of variable fractional order. The problem is then stated in Sect. 4.2.1, consisting of the variational functional

$$\mathcal{J}(x, T) = \int_a^T L\left(t, x(t), {}^C D_\gamma^{\alpha(\cdot,\cdot),\beta(\cdot,\cdot)} x(t)\right) dt + \phi(T, x(T)),$$

where ${}^C D_\gamma^{\alpha(\cdot,\cdot),\beta(\cdot,\cdot)} x(t)$ stands for the combined Caputo fractional derivative of variable fractional order (Definition 33), subject to the boundary condition $x(a) = x_a$.

In this problem, we do not only assume that $x(T)$ is free, but the endpoint T is also variable. Therefore, we are interested in finding an optimal curve $x(\cdot)$ and also the endpoint of the variational integral, denoted in the sequel by T.

We begin by proving in Sect. 4.2.1 necessary optimality conditions that every extremizer (x, T) must satisfy. The main results of this section provide necessary optimality conditions of Euler–Lagrange type, described by fractional differential equations of variable order and different transversality optimality conditions (Theorems 45 and 46). Some particular cases of interest are considered in Sect. 4.2.2. We end with two illustrative examples (Sect. 4.2.3).

4.2.1 Necessary Optimality Conditions

Let $\alpha, \beta : [a, b]^2 \to (0, 1)$ be two functions. Let D denote the set

$$D = \left\{(x, t) \in C^1([a, b]) \times [a, b] : {}^C D_\gamma^{\alpha(\cdot,\cdot),\beta(\cdot,\cdot)} x \in C([a, b])\right\},\qquad (4.6)$$

endowed with the norm $\|(\cdot, \cdot)\|$ defined on the linear space $C^1([a, b]) \times \mathbb{R}$ by

$$\|(x, t)\| := \max_{a \le t \le b} |x(t)| + \max_{a \le t \le b} \left|{}^C D_\gamma^{\alpha(\cdot,\cdot),\beta(\cdot,\cdot)} x(t)\right| + |t|.$$

Definition 39 We say that $(x^\star, T^\star) \in D$ is a local minimizer to the functional $\mathcal{J} : D \to \mathbb{R}$ if there exists some $\epsilon > 0$ such that

$$\forall (x, T) \in D : \quad \|(x^\star, T^\star) - (x, T)\| < \epsilon \Rightarrow \mathcal{J}(x^\star, T^\star) \le \mathcal{J}(x, T).$$

Along the work, we denote by $\partial_i z, i \in \{1, 2, 3\}$, the partial derivative of a function $z : \mathbb{R}^3 \to \mathbb{R}$ with respect to its ith argument, and by L a differentiable Lagrangian $L : [a, b] \times \mathbb{R}^2 \to \mathbb{R}$.

Consider the following problem of the calculus of variations:

Problem 1 Find the local minimizers of the functional $\mathcal{J} : D \to \mathbb{R}$, with

$$\mathcal{J}(x, T) = \int_a^T L\left(t, x(t), {}^C D_\gamma^{\alpha(\cdot,\cdot),\beta(\cdot,\cdot)} x(t)\right) dt + \phi(T, x(T)), \qquad (4.7)$$

over all $(x, T) \in D$ satisfying the boundary condition $x(a) = x_a$, for a fixed $x_a \in \mathbb{R}$. The terminal time T and terminal state $x(T)$ are free.

The terminal cost function $\phi : [a, b] \times \mathbb{R} \to \mathbb{R}$ is at least of class C^1.

For simplicity of notation, we introduce the operator $[\cdot]_\gamma^{\alpha,\beta}$ defined by

$$[x]_\gamma^{\alpha,\beta}(t) = \left(t, x(t), {}^C D_\gamma^{\alpha(\cdot,\cdot),\beta(\cdot,\cdot)} x(t)\right). \qquad (4.8)$$

With the new notation, one can write (4.7) simply as

$$\mathcal{J}(x, T) = \int_a^T L[x]_\gamma^{\alpha,\beta}(t) dt + \phi(T, x(T)).$$

The next theorem gives fractional necessary optimality conditions to Problem 1.

Theorem 45 *Suppose that (x, T) is a local minimizer to the functional (4.7) on D. Then, (x, T) satisfies the fractional Euler–Lagrange equations*

$$\partial_2 L[x]_\gamma^{\alpha,\beta}(t) + D_{\overline{\gamma}}^{\beta(\cdot,\cdot),\alpha(\cdot,\cdot)} \partial_3 L[x]_\gamma^{\alpha,\beta}(t) = 0, \qquad (4.9)$$

on the interval $[a, T]$ and

$$\gamma_2 \left({}_a D_t^{\beta(\cdot,\cdot)} \partial_3 L[x]_\gamma^{\alpha,\beta}(t) - {}_T D_t^{\beta(\cdot,\cdot)} \partial_3 L[x]_\gamma^{\alpha,\beta}(t)\right) = 0, \qquad (4.10)$$

on the interval $[T, b]$. *Moreover,* (x, T) *satisfies the transversality conditions*

$$\begin{cases} L[x]_\gamma^{\alpha,\beta}(T) + \partial_1\phi(T, x(T)) + \partial_2\phi(T, x(T))x'(T) = 0, \\ \left[\gamma_1\,{}_tI_T^{1-\alpha(\cdot,\cdot)}\partial_3L[x]_\gamma^{\alpha,\beta}(t) - \gamma_2\,{}_TI_t^{1-\beta(\cdot,\cdot)}\partial_3L[x]_\gamma^{\alpha,\beta}(t)\right]_{t=T} + \partial_2\phi(T, x(T)) = 0, \\ \gamma_2\left[{}_TI_t^{1-\beta(\cdot,\cdot)}\partial_3L[x]_\gamma^{\alpha,\beta}(t) - {}_aI_t^{1-\beta(\cdot,\cdot)}\partial_3L[x]_\gamma^{\alpha,\beta}(t)\right]_{t=b} = 0. \end{cases}$$

$$(4.11)$$

Proof Let (x, T) be a solution to the problem and $(x + \epsilon h, T + \epsilon\Delta T)$ be an admissible variation, where $h \in C^1([a, b]; \mathbb{R})$ is a perturbing curve, $\Delta T \in \mathbb{R}$ represents an arbitrarily chosen small change in T and $\epsilon \in \mathbb{R}$ represents a small number. The constraint $x(a) = x_a$ implies that all admissible variations must fulfill the condition $h(a) = 0$. Define $j(\cdot)$ on a neighborhood of zero by

$$\begin{aligned} j(\epsilon) &= \mathcal{J}(x + \epsilon h, T + \epsilon\Delta T) \\ &= \int_a^{T+\epsilon\Delta T} L[x + \epsilon h]_\gamma^{\alpha,\beta}(t)\,dt + \phi\left(T + \epsilon\Delta T, (x + \epsilon h)(T + \epsilon\Delta T)\right). \end{aligned}$$

The derivative $j'(\epsilon)$ is

$$\begin{aligned} \int_a^{T+\epsilon\Delta T} &\left(\partial_2 L[x + \epsilon h]_\gamma^{\alpha,\beta}(t)h(t) + \partial_3 L[x + \epsilon h]_\gamma^{\alpha,\beta}(t){}^C D_\gamma^{\alpha(\cdot,\cdot),\beta(\cdot,\cdot)}h(t)\right) dt \\ &+ L[x + \epsilon h]_\gamma^{\alpha,\beta}(T + \epsilon\Delta T)\Delta T + \partial_1\phi\left(T + \epsilon\Delta T, (x + \epsilon h)(T + \epsilon\Delta T)\right)\Delta T \\ &+ \partial_2\phi\left(T + \epsilon\Delta T, (x + \epsilon h)(T + \epsilon\Delta T)\right)\left[(x + \epsilon h)(T + \epsilon\Delta T)\right]'. \end{aligned}$$

Considering the differentiability properties of j, a necessary condition for (x, T) to be a local extremizer is given by $j'(\epsilon)\big|_{\epsilon=0} = 0$, that is,

$$\int_a^T \left(\partial_2 L[x]_\gamma^{\alpha,\beta}(t)h(t) + \partial_3 L[x]_\gamma^{\alpha,\beta}(t){}^C D_\gamma^{\alpha(\cdot,\cdot),\beta(\cdot,\cdot)}h(t)\right) dt + L[x]_\gamma^{\alpha,\beta}(T)\Delta T$$
$$+ \partial_1\phi\left(T, x(T)\right)\Delta T + \partial_2\phi(T, x(T))\left[h(T) + x'(T)\Delta T\right] = 0. \quad (4.12)$$

The second addend of the integral function (4.12),

$$\int_a^T \partial_3 L[x]_\gamma^{\alpha,\beta}(t){}^C D_\gamma^{\alpha(\cdot,\cdot),\beta(\cdot,\cdot)}h(t)dt, \quad (4.13)$$

can be written, using the definition of combined Caputo fractional derivative, as

$$\begin{aligned} \int_a^T &\partial_3 L[x]_\gamma^{\alpha,\beta}(t){}^C D_\gamma^{\alpha(\cdot,\cdot),\beta(\cdot,\cdot)}h(t)dt \\ &= \int_a^T \partial_3 L[x]_\gamma^{\alpha,\beta}(t)\left[\gamma_1\,{}_a^C D_t^{\alpha(\cdot,\cdot)}h(t) + \gamma_2\,{}_t^C D_b^{\beta(\cdot,\cdot)}h(t)\right] dt \end{aligned}$$

$$= \gamma_1 \int_a^T \partial_3 L[x]_\gamma^{\alpha,\beta}(t) {}_a^C D_t^{\alpha(\cdot,\cdot)} h(t) dt$$

$$+ \gamma_2 \left[\int_a^b \partial_3 L[x]_\gamma^{\alpha,\beta}(t) {}_t^C D_b^{\beta(\cdot,\cdot)} h(t) dt - \int_T^b \partial_3 L[x]_\gamma^{\alpha,\beta}(t) {}_t^C D_b^{\beta(\cdot,\cdot)} h(t) dt \right].$$

Integrating by parts (see Theorem 13), and since $h(a) = 0$, the term (4.13) can be written as

$$\gamma_1 \left[\int_a^T h(t) {}_t D_T^{\alpha(\cdot,\cdot)} \partial_3 L[x]_\gamma^{\alpha,\beta}(t) dt + \left[h(t) {}_t I_T^{1-\alpha(\cdot,\cdot)} \partial_3 L[x]_\gamma^{\alpha,\beta}(t) \right]_{t=T} \right]$$

$$+ \gamma_2 \left[\int_a^b h(t) {}_a D_t^{\beta(\cdot,\cdot)} \partial_3 L[x]_\gamma^{\alpha,\beta}(t) dt - \left[h(t) {}_a I_t^{1-\beta(\cdot,\cdot)} \partial_3 L[x]_\gamma^{\alpha,\beta}(t) \right]_{t=b} \right.$$

$$- \left(\int_T^b h(t) {}_T D_t^{\beta(\cdot,\cdot)} \partial_3 L[x]_\gamma^{\alpha,\beta}(t) dt - \left[h(t) {}_T I_t^{1-\beta(\cdot,\cdot)} \partial_3 L[x]_\gamma^{\alpha,\beta}(t) \right]_{t=b} \right.$$

$$\left. \left. + \left[h(t) {}_T I_t^{1-\beta(\cdot,\cdot)} \partial_3 L[x]_\gamma^{\alpha,\beta}(t) \right]_{t=T} \right) \right].$$

Unfolding these integrals, and considering the fractional operator $D_{\bar\gamma}^{\beta(\cdot,\cdot),\alpha(\cdot,\cdot)}$ with $\bar\gamma = (\gamma_2, \gamma_1)$, then (4.13) is equivalent to

$$\int_a^T h(t) D_{\bar\gamma}^{\beta(\cdot,\cdot),\alpha(\cdot,\cdot)} \partial_3 L[x]_\gamma^{\alpha,\beta}(t) dt$$

$$+ \int_T^b \gamma_2 h(t) \left[{}_a D_t^{\beta(\cdot,\cdot)} \partial_3 L[x]_\gamma^{\alpha,\beta}(t) - {}_T D_t^{\beta(\cdot,\cdot)} \partial_3 L[x]_\gamma^{\alpha,\beta}(t) \right] dt$$

$$+ \left[h(t) \left(\gamma_1 {}_t I_T^{1-\alpha(\cdot,\cdot)} \partial_3 L[x]_\gamma^{\alpha,\beta}(t) - \gamma_2 {}_T I_t^{1-\beta(\cdot,\cdot)} \partial_3 L[x]_\gamma^{\alpha,\beta}(t) \right) \right]_{t=T}$$

$$+ \left[h(t) \gamma_2 \left({}_T I_t^{1-\beta(\cdot,\cdot)} \partial_3 L[x]_\gamma^{\alpha,\beta}(t) - {}_a I_t^{1-\beta(\cdot,\cdot)} \partial_3 L[x]_\gamma^{\alpha,\beta}(t) \right) \right]_{t=b}.$$

Substituting these relations into Eq. (4.12), we obtain

$$0 = \int_a^T h(t) \left[\partial_2 L[x]_\gamma^{\alpha,\beta}(t) + D_{\bar\gamma}^{\beta(\cdot,\cdot),\alpha(\cdot,\cdot)} \partial_3 L[x]_\gamma^{\alpha,\beta}(t) \right] dt$$

$$+ \int_T^b \gamma_2 h(t) \left[{}_a D_t^{\beta(\cdot,\cdot)} \partial_3 L[x]_\gamma^{\alpha,\beta}(t) - {}_T D_t^{\beta(\cdot,\cdot)} \partial_3 L[x]_\gamma^{\alpha,\beta}(t) \right] dt$$

$$+ h(T) \left[\gamma_1 {}_t I_T^{1-\alpha(\cdot,\cdot)} \partial_3 L[x]_\gamma^{\alpha,\beta}(t) - \gamma_2 {}_T I_t^{1-\beta(\cdot,\cdot)} \partial_3 L[x]_\gamma^{\alpha,\beta}(t) \right. \tag{4.14}$$

$$\left. + \partial_2 \phi(t, x(t)) \right]_{t=T}$$

$$+ \Delta T \left[L[x]_\gamma^{\alpha,\beta}(t) + \partial_1 \phi(t, x(t)) + \partial_2 \phi(t, x(t)) x'(t) \right]_{t=T}$$

$$+ h(b) \left[\gamma_2 \left({}_T I_t^{1-\beta(\cdot,\cdot)} \partial_3 L[x]_\gamma^{\alpha,\beta}(t) - {}_a I_t^{1-\beta(\cdot,\cdot)} \partial_3 L[x]_\gamma^{\alpha,\beta}(t) \right) \right]_{t=b}.$$

As h and $\triangle T$ are arbitrary, we can choose $\triangle T = 0$ and $h(t) = 0$, for all $t \in [T, b]$, but h is arbitrary in $t \in [a, T)$. Then, for all $t \in [a, T]$, we obtain the first necessary condition (4.9):

$$\partial_2 L[x]_\gamma^{\alpha,\beta}(t) + D_\gamma^{\beta(\cdot,\cdot),\alpha(\cdot,\cdot)}\partial_3 L[x]_\gamma^{\alpha,\beta}(t) = 0.$$

Analogously, considering $\triangle T = 0, h(t) = 0$, for all $t \in [a, T] \cup \{b\}$, and h arbitrary on (T, b), we obtain the second necessary condition (4.10):

$$\gamma_2 \left({_a}D_t^{\beta(\cdot,\cdot)}\partial_3 L[x]_\gamma^{\alpha,\beta}(t) - {_T}D_t^{\beta(\cdot,\cdot)}\partial_3 L[x]_\gamma^{\alpha,\beta}(t) \right) = 0.$$

As (x, T) is a solution to the necessary conditions (4.9) and (4.10), then Eq. (4.14) takes the form

$$
\begin{aligned}
0 = h(T) & \left[\gamma_1 \, {_t}I_T^{1-\alpha(\cdot,\cdot)}\partial_3 L[x]_\gamma^{\alpha,\beta}(t) - \gamma_2 \, {_T}I_t^{1-\beta(\cdot,\cdot)}\partial_3 L[x]_\gamma^{\alpha,\beta}(t) + \partial_2\phi(t, x(t)) \right]_{t=T} \\
& + \triangle T \left[L[x]_\gamma^{\alpha,\beta}(t) + \partial_1\phi(t, x(t)) + \partial_2\phi(t, x(t))x'(t) \right]_{t=T} \\
& + h(b) \left[\gamma_2 \left({_T}I_t^{1-\beta(\cdot,\cdot)}\partial_3 L[x]_\gamma^{\alpha,\beta}(t) - {_a}I_t^{1-\beta(\cdot,\cdot)}\partial_3 L[x]_\gamma^{\alpha,\beta}(t) \right) \right]_{t=b}.
\end{aligned}
$$
(4.15)

Transversality conditions (4.11) are obtained for appropriate choices of variations.

In the next theorem, considering the same Problem 1, we rewrite the transversality conditions (4.11) in terms of the increment $\triangle T$ on time and on the consequent increment $\triangle x_T$ on x, given by

$$\triangle x_T = (x + h)(T + \triangle T) - x(T).$$
(4.16)

Theorem 46 *Let (x, T) be a local minimizer to the functional (4.7) on D. Then, the fractional Euler–Lagrange equations (4.9) and (4.10) are satisfied together with the following transversality conditions:*

$$
\begin{cases}
L[x]_\gamma^{\alpha,\beta}(T) + \partial_1\phi(T, x(T)) \\
\quad + x'(T)\left[\gamma_2{_T}I_t^{1-\beta(\cdot,\cdot)}\partial_3 L[x]_\gamma^{\alpha,\beta}(t) - \gamma_1{_t}I_T^{1-\alpha(\cdot,\cdot)}\partial_3 L[x]_\gamma^{\alpha,\beta}(t) \right]_{t=T} = 0, \\
\left[\gamma_1 \, {_t}I_T^{1-\alpha(\cdot,\cdot)}\partial_3 L[x]_\gamma^{\alpha,\beta}(t) - \gamma_2 \, {_T}I_t^{1-\beta(\cdot,\cdot)}\partial_3 L[x]_\gamma^{\alpha,\beta}(t) \right]_{t=T} + \partial_2\phi(T, x(T)) = 0, \\
\gamma_2 \left[{_T}I_t^{1-\beta(\cdot,\cdot)}\partial_3 L[x]_\gamma^{\alpha,\beta}(t) - {_a}I_t^{1-\beta(\cdot,\cdot)}\partial_3 L[x]_\gamma^{\alpha,\beta}(t) \right]_{t=b} = 0.
\end{cases}
$$
(4.17)

Proof The Euler–Lagrange equations are deduced following similar arguments as the ones presented in Theorem 45. We now focus our attention on the proof of the transversality conditions. Using Taylor's expansion up to first order for a small $\triangle T$, and restricting the set of variations to those for which $h'(T) = 0$, we obtain

$$(x + h)(T + \triangle T) = (x + h)(T) + x'(T)\triangle T + O(\triangle T)^2.$$

Rearranging the relation (4.16) allows us to express $h(T)$ in terms of ΔT and Δx_T:

$$h(T) = \Delta x_T - x'(T)\Delta T + O(\Delta T)^2.$$

Substitution of this expression into (4.15) gives us

$$
\begin{aligned}
0 = \Delta x_T & \left[\gamma_{1\, t} I_T^{1-\alpha(\cdot,\cdot)} \partial_3 L[x]_\gamma^{\alpha,\beta}(t) - \gamma_{2\, T} I_t^{1-\beta(\cdot,\cdot)} \partial_3 L[x]_\gamma^{\alpha,\beta}(t) + \partial_2\phi(t,x(t)) \right]_{t=T} \\
& + \Delta T \left[L[x]_\gamma^{\alpha,\beta}(t) + \partial_1\phi(t,x(t)) \right. \\
& \quad \left. -x'(t) \left(\gamma_{1\, t} I_T^{1-\alpha(\cdot,\cdot)} \partial_3 L[x]_\gamma^{\alpha,\beta}(t) - \gamma_{2\, T} I_t^{1-\beta(\cdot,\cdot)} \partial_3 L[x]_\gamma^{\alpha,\beta}(t) \right) \right]_{t=T} \\
& + h(b) \left[\gamma_2 \left({}_T I_t^{1-\beta(\cdot,\cdot)} \partial_3 L[x]_\gamma^{\alpha,\beta}(t) - {}_a I_t^{1-\beta(\cdot,\cdot)} \partial_3 L[x]_\gamma^{\alpha,\beta}(t) \right) \right]_{t=b} + O(\Delta T)^2.
\end{aligned}
$$

Transversality conditions (4.17) are obtained using appropriate choices of variations.

4.2.2 Particular Cases

Now, we specify our results to three particular cases of variable terminal points.

Vertical Terminal Line

This case involves a fixed upper bound T. Thus, $\Delta T = 0$ and, consequently, the second term in (4.15) drops out. Since Δx_T is arbitrary, we obtain the following transversality conditions: if $T < b$, then

$$
\begin{cases}
\left[\gamma_{1\, t} I_T^{1-\alpha(\cdot,\cdot)} \partial_3 L[x]_\gamma^{\alpha,\beta}(t) - \gamma_{2\, T} I_t^{1-\beta(\cdot,\cdot)} \partial_3 L[x]_\gamma^{\alpha,\beta}(t) \right]_{t=T} + \partial_2\phi(T, x(T)) = 0, \\
\gamma_2 \left[{}_T I_t^{1-\beta(\cdot,\cdot)} \partial_3 L[x]_\gamma^{\alpha,\beta}(t) - {}_a I_t^{1-\beta(\cdot,\cdot)} \partial_3 L[x]_\gamma^{\alpha,\beta}(t) \right]_{t=b} = 0;
\end{cases}
$$

if $T = b$, then $\Delta x_T = h(b)$, and the transversality conditions reduce to

$$
\left[\gamma_{1\, t} I_b^{1-\alpha(\cdot,\cdot)} \partial_3 L[x]_\gamma^{\alpha,\beta}(t) - \gamma_{2\, a} I_t^{1-\beta(\cdot,\cdot)} \partial_3 L[x]_\gamma^{\alpha,\beta}(t) \right]_{t=b} + \partial_2\phi(b, x(b)) = 0.
$$

Horizontal Terminal Line

In this situation, we have $\Delta x_T = 0$ but ΔT is arbitrary. Thus, the transversality conditions are

$$
\begin{cases}
L[x]_\gamma^{\alpha,\beta}(T) + \partial_1\phi(T, x(T)) \\
\quad +x'(T) \left[\gamma_{2\, T} I_t^{1-\beta(\cdot,\cdot)} \partial_3 L[x]_\gamma^{\alpha,\beta}(t) - \gamma_{1\, t} I_T^{1-\alpha(\cdot,\cdot)} \partial_3 L[x]_\gamma^{\alpha,\beta}(t) \right]_{t=T} = 0, \\
\gamma_2 \left[{}_T I_t^{1-\beta(\cdot,\cdot)} \partial_3 L[x]_\gamma^{\alpha,\beta}(t) - {}_a I_t^{1-\beta(\cdot,\cdot)} \partial_3 L[x]_\gamma^{\alpha,\beta}(t) \right]_{t=b} = 0.
\end{cases}
$$

Terminal Curve

Now, the terminal point is described by a given curve $\psi : C^1([a, b]) \to \mathbb{R}$, in the sense that $x(T) = \psi(T)$. From Taylor's formula, for a small arbitrary ΔT, one has

$$\Delta x(T) = \psi'(T)\Delta T + O(\Delta T)^2.$$

Hence, the transversality conditions are presented in the form

$$
\begin{cases}
L[x]_\gamma^{\alpha,\beta}(T) + \partial_1\phi(T, x(T)) + \partial_2\phi(T, x(T))\psi'(T) + \left(x'(T) - \psi'(T)\right) \\
\quad \times \left[\gamma_2\, _TI_t^{1-\beta(\cdot,\cdot)}\partial_3 L[x]_\gamma^{\alpha,\beta}(t) - \gamma_1\, _tI_T^{1-\alpha(\cdot,\cdot)}\partial_3 L[x]_\gamma^{\alpha,\beta}(t)\right]_{t=T} = 0, \\
\gamma_2\left[_TI_t^{1-\beta(\cdot,\cdot)}\partial_3 L[x]_\gamma^{\alpha,\beta}(t) - {}_aI_t^{1-\beta(\cdot,\cdot)}\partial_3 L[x]_\gamma^{\alpha,\beta}(t)\right]_{t=b} = 0.
\end{cases}
$$

4.2.3 Examples

In this section, we show two examples to illustrate the new results. Let $\alpha(t, \tau) = \alpha(t)$ and $\beta(t, \tau) = \beta(\tau)$ be two functions depending on a variable t and τ only, respectively. Consider the following fractional variational problem: to minimize the functional

$$
\mathcal{J}(x, T) = \int_0^T \Bigg[2\alpha(t) - 1
$$
$$
+ \left({}^C D_\gamma^{\alpha(\cdot),\beta(\cdot)}x(t) - \frac{t^{1-\alpha(t)}}{2\Gamma(2 - \alpha(t))} - \frac{(10 - t)^{1-\beta(t)}}{2\Gamma(2 - \beta(t))}\right)^2\Bigg]dt
$$

for $t \in [0, 10]$, subject to the initial condition $x(0) = 0$ and where $\gamma = (\gamma_1, \gamma_2) = (1/2, 1/2)$. Simple computations show that for $\overline{x}(t) = t$, with $t \in [0, 10]$, we have

$$
{}^C D_\gamma^{\alpha(\cdot),\beta(\cdot)}\overline{x}(t) = \frac{t^{1-\alpha(t)}}{2\Gamma(2 - \alpha(t))} + \frac{(10 - t)^{1-\beta(t)}}{2\Gamma(2 - \beta(t))}.
$$

For $\overline{x}(t) = t$, the functional reduces to

$$
\mathcal{J}(\overline{x}, T) = \int_0^T (2\alpha(t) - 1)\, dt.
$$

In order to determine the optimal time T, we have to solve the equation $2\alpha(T) = 1$. For example, let $\alpha(t) = t^2/2$. In this case, since $\mathcal{J}(x, T) \geq -2/3$ for all pairs (x, T) and $\mathcal{J}(\overline{x}, 1) = -2/3$, we conclude that the (global) minimum value of the functional is $-2/3$, obtained for $\overline{x}$ and $T = 1$. It is obvious that the two Euler–Lagrange equations (4.9) and (4.10) are satisfied when $x = \overline{x}$, since

$$\partial_3 L[\overline{x}]_\gamma^{\alpha,\beta}(t) = 0 \quad \text{for all} \quad t \in [0, 10].$$

Using this relation, together with

$$L[\overline{x}]_\gamma^{\alpha,\beta}(1) = 0,$$

the transversality conditions (4.11) are also verified.

For our last example, consider the functional

$$\mathcal{J}(x, T) = \int_0^T \left[2\alpha(t) - 1 \right.$$
$$\left. + \left({}^C D_\gamma^{\alpha(\cdot),\beta(\cdot)} x(t) - \frac{t^{1-\alpha(t)}}{2\Gamma(2-\alpha(t))} - \frac{(10-t)^{1-\beta(t)}}{2\Gamma(2-\beta(t))} \right)^3 \right] dt,$$

where the remaining assumptions and conditions are as in the previous example. For this case, $\overline{x}(t) = t$ and $T = 1$ still satisfy the necessary optimality conditions. However, we cannot assure that $(\overline{x}, 1)$ is a local minimizer to the problem.

4.3 Higher-Order Variational Problems

In this section, we intend to generalize the results obtained in Sect. 4.2 by considering higher-order variational problems with a Lagrangian depending on a higher-order combined Caputo derivative of variable fractional order, defined by

$$^C D_{\gamma^n}^{\alpha_n(\cdot,\cdot),\beta_n(\cdot,\cdot)} x(t) = \gamma_1^n \, {}_a^C D_t^{\alpha_n(\cdot,\cdot)} x(t) + \gamma_2^n \, {}_t^C D_b^{\beta_n(\cdot,\cdot)} x(t),$$

subject to boundary conditions at the initial time $t = a$.

In Sect. 4.3.1, we obtain higher-order Euler–Lagrange equations and transversality conditions for the generalized variational problem with a Lagrangian depending on a combined Caputo fractional derivative of variable fractional order (Theorems 47 and 49).

One illustrative example is discussed in Sect. 4.3.2.

4.3.1 Necessary Optimality Conditions

Let $n \in \mathbb{N}$ and $x : [a, b] \to \mathbb{R}$ be a function of class C^n. The fractional order is a continuous function of two variables, $\alpha_n : [a, b]^2 \to (n - 1, n)$.

Let D denote the linear subspace of $C^n([a, b]) \times [a, b]$ such that the fractional derivative of x, $^C D_{\gamma_i}^{\alpha_i(\cdot,\cdot),\beta_i(\cdot,\cdot)} x(t)$, exists and is continuous on the interval $[a, b]$ for all $i \in \{1, \ldots, n\}$. We endow D with the norm

$$\|(x, t)\| = \max_{a \leq t \leq b} |x(t)| + \max_{a \leq t \leq b} \sum_{i=1}^{n} \left| {}^{C}D_{\gamma^i}^{\alpha_i(\cdot,\cdot),\beta_i(\cdot,\cdot)} x(t) \right| + |t|.$$

Consider the following higher-order problem of the calculus of variations:

Problem 2 Minimize functional $\mathcal{J} : D \to \mathbb{R}$, where

$$\mathcal{J}(x, T) = \int_{a}^{T} L\left(t, x(t), {}^{C}D_{\gamma^1}^{\alpha_1(\cdot,\cdot),\beta_1(\cdot,\cdot)} x(t), \ldots, {}^{C}D_{\gamma^n}^{\alpha_n(\cdot,\cdot),\beta_n(\cdot,\cdot)} x(t)\right) dt$$

$$+ \phi(T, x(T)), \quad (4.18)$$

over all $(x, T) \in D$ subject to boundary conditions

$$x(a) = x_a, \quad x^{(i)}(a) = x_a^i, \quad \forall i \in \{1, \ldots, n-1\},$$

for fixed $x_a, x_a^1, \ldots, x_a^{n-1} \in \mathbb{R}$. Here, the terminal time T and terminal state $x(T)$ are both free. For all $i \in \{1, \ldots, n\}$, $\alpha_i, \beta_i\left([a, b]^2\right) \subseteq (i-1, i)$ and $\gamma^i = \left(\gamma_1^i, \gamma_2^i\right)$ is a vector. The terminal cost function $\phi : [a, b] \times \mathbb{R} \to \mathbb{R}$ is at least of class C^1.

For simplicity of notation, we introduce the operator $[\cdot]_{\gamma}^{\alpha,\beta}$ defined by

$$[x]_{\gamma}^{\alpha,\beta}(t) = \left(t, x(t), {}^{C}D_{\gamma^1}^{\alpha_1(\cdot,\cdot),\beta_1(\cdot,\cdot)} x(t), \ldots, {}^{C}D_{\gamma^n}^{\alpha_n(\cdot,\cdot),\beta_n(\cdot,\cdot)} x(t)\right).$$

We assume that the Lagrangian $L : [a, b] \times \mathbb{R}^{n+1} \to \mathbb{R}$ is a function of class C^1. Along the work, we denote by $\partial_i L$, $i \in \{1, \ldots, n+2\}$, the partial derivative of the Lagrangian L with respect to its ith argument.

Now, we can rewrite functional (4.18) as

$$\mathcal{J}(x, T) = \int_{a}^{T} L[x]_{\gamma}^{\alpha,\beta}(t) dt + \phi(T, x(T)). \quad (4.19)$$

In the previous section, we obtained fractional necessary optimality conditions that every local minimizer of functional $\mathcal{J}$, with $n = 1$, must fulfill. Here, we generalize those results to arbitrary values of n, $n \in \mathbb{N}$. Necessary optimality conditions for Problem 2 are presented next.

Theorem 47 *Suppose that (x, T) gives a minimum to functional (4.19) on D. Then, (x, T) satisfies the following fractional Euler–Lagrange equations:*

$$\partial_2 L[x]_{\gamma}^{\alpha,\beta}(t) + \sum_{i=1}^{n} D_{\gamma^i}^{\beta_i(\cdot,\cdot),\alpha_i(\cdot,\cdot)} \partial_{i+2} L[x]_{\gamma}^{\alpha,\beta}(t) = 0, \quad (4.20)$$

on the interval $[a, T]$ and

$$\sum_{i=1}^{n} \gamma_2^i \left({}_aD_t^{\beta_i(\cdot,\cdot)} \partial_{i+2} L[x]_\gamma^{\alpha,\beta}(t) - {}_TD_t^{\beta_i(\cdot,\cdot)} \partial_{i+2} L[x]_\gamma^{\alpha,\beta}(t) \right) = 0, \qquad (4.21)$$

on the interval $[T, b]$. *Moreover,* (x, T) *satisfies the following transversality conditions:*

$$
\begin{cases}
L[x]_\gamma^{\alpha, \beta}(T) + \partial_1 \phi(T, x(T)) + \partial_2 \phi(T, x(T)) x'(T) = 0, \\
\sum_{i=1}^{n} \left[\gamma_1^i (-1)^{i-1} \frac{d^{i-1}}{dt^{i-1}} {}_t I_T^{i-\alpha_i(\cdot,\cdot)} \partial_{i+2} L[x]_\gamma^{\alpha,\beta}(t) \right. \\
\left. \qquad - \gamma_2^i \frac{d^{i-1}}{dt^{i-1}} {}_T I_t^{i-\beta_i(\cdot,\cdot)} \partial_{i+2} L[x]_\gamma^{\alpha,\beta}(t) \right]_{t=T} + \partial_2 \phi(T, x(T)) = 0, \\
\sum_{i=j+1}^{n} \left[\gamma_1^i (-1)^{i-1-j} \frac{d^{i-1-j}}{dt^{i-1-j}} {}_t I_T^{i-\alpha_i(\cdot,\cdot)} \partial_{i+2} L[x]_\gamma^{\alpha,\beta}(t) \right. \\
\left. \qquad + \gamma_2^i (-1)^{j+1} \frac{d^{i-1-j}}{dt^{i-1-j}} {}_T I_t^{i-\beta_i(\cdot,\cdot)} \partial_{i+2} L[x]_\gamma^{\alpha,\beta}(t) \right]_{t=T} = 0, \quad \forall j = 1, \ldots, n-1, \\
\sum_{i=j+1}^{n} \left[\gamma_2^i (-1)^{j+1} \left[\frac{d^{i-1-j}}{dt^{i-1-j}} {}_a I_t^{i-\beta_i(\cdot,\cdot)} \partial_{i+2} L[x]_\gamma^{\alpha,\beta}(t) \right. \right. \\
\left. \left. \qquad - \frac{d^{i-1-j}}{dt^{i-1-j}} {}_T I_t^{i-\beta_i(\cdot,\cdot)} \partial_{i+2} L[x]_\gamma^{\alpha,\beta}(t) \right] \right]_{t=b} = 0, \quad \forall j = 0, \ldots, n-1.
\end{cases}
$$
$$(4.22)$$

Proof The proof is an extension of the one used in Theorem 45. Let $h \in C^n([a, b]; \mathbb{R})$ be a perturbing curve and $\Delta T \in \mathbb{R}$ an arbitrarily chosen small change in T. For a small number $\epsilon \in \mathbb{R}$, if (x, T) is a solution to the problem, we consider an admissible variation of (x, T) of the form $(x + \epsilon h, T + \epsilon \Delta T)$, and then, by the minimum condition, we have that

$$\mathcal{J}(x, T) \leq \mathcal{J}(x + \epsilon h, T + \epsilon \Delta T).$$

The constraints $x^{(i)}(a) = x_a^{(i)}$ imply that all admissible variations must fulfill the conditions $h^{(i)}(a) = 0$, for all $i = 0, \ldots, n-1$. We define function $j(\cdot)$ on a neighborhood of zero by

$$
\begin{aligned}
j(\epsilon) &= \mathcal{J}(x + \epsilon h, T + \epsilon \Delta T) \\
&= \int_a^{T+\epsilon \Delta T} L[x + \epsilon h]_\gamma^{\alpha,\beta}(t) \, dt + \phi\left(T + \epsilon \Delta T, (x + \epsilon h)(T + \epsilon \Delta T)\right).
\end{aligned}
$$

The derivative $j'(\epsilon)$ is given by the expression

$$\int_a^{T+\epsilon \Delta T} \left(\partial_2 L[x + \epsilon h]_\gamma^{\alpha,\beta}(t) h(t) + \sum_{i=1}^{n} \partial_{i+2} L[x + \epsilon h]_\gamma^{\alpha,\beta}(t) {}^C D_{\gamma^i}^{\alpha_i(\cdot,\cdot), \beta_i(\cdot,\cdot)} h(t) \right) dt$$

$$+ L[x + \epsilon h]_\gamma^{\alpha,\beta}(T + \epsilon\Delta T)\Delta T + \partial_1\phi(T + \epsilon\Delta T, (x + \epsilon h)(T + \epsilon\Delta T))\ \Delta T$$
$$+ \partial_2\phi(T + \epsilon\Delta T, (x + \epsilon h)(T + \epsilon\Delta T))\ (x + \epsilon h)'(T + \epsilon\Delta T).$$

Hence, by Fermat's theorem, a necessary condition for (x, T) to be a local minimizer of j is given by $j'(0) = 0$, that is,

$$\int_a^T \left(\partial_2 L[x]_\gamma^{\alpha,\beta}(t)h(t) + \sum_{i=1}^n \partial_{i+2}L[x]_\gamma^{\alpha,\beta}(t)^C D_{\gamma^i}^{\alpha_i(\cdot,\cdot),\beta_i(\cdot,\cdot)}h(t) \right) dt$$

$$+ L[x]_\gamma^{\alpha,\beta}(T)\Delta T + \partial_1\phi(T, x(T))\,\Delta T + \partial_2\phi(T, x(T))\left[h(t) + x'(T)\Delta T\right] = 0. \tag{4.23}$$

Considering the second addend of the integral function (4.23), for $i = 1$, we get

$$\int_a^T \partial_3 L[x]_\gamma^{\alpha,\beta}(t)^C D_{\gamma^1}^{\alpha_1(\cdot,\cdot),\beta_1(\cdot,\cdot)}h(t)dt$$

$$= \int_a^T \partial_3 L[x]_\gamma^{\alpha,\beta}(t)\left[\gamma_1^1{}_a^C D_t^{\alpha_1(\cdot,\cdot)}h(t) + \gamma_2^1{}_t^C D_b^{\beta_1(\cdot,\cdot)}h(t)\right]dt$$

$$= \gamma_1^1 \int_a^T \partial_3 L[x]_\gamma^{\alpha,\beta}(t)_a^C D_t^{\alpha_1(\cdot,\cdot)}h(t)dt$$

$$+ \gamma_2^1\left[\int_a^b \partial_3 L[x]_\gamma^{\alpha,\beta}(t)_t^C D_b^{\beta_1(\cdot,\cdot)}h(t)dt - \int_T^b \partial_3 L[x]_\gamma^{\alpha,\beta}(t)_t^C D_b^{\beta_1(\cdot,\cdot)}h(t)dt\right].$$

Integrating by parts (see Theorem 42), and since $h(a) = 0$, we obtain that

$$\gamma_1^1\left[\int_a^T h(t)_t D_T^{\alpha_1(\cdot,\cdot)}\partial_3 L[x]_\gamma^{\alpha,\beta}(t)dt + \left[h(t)_t I_T^{1-\alpha_1(\cdot,\cdot)}\partial_3 L[x]_\gamma^{\alpha,\beta}(t)\right]_{t=T}\right]$$

$$+ \gamma_2^1\left[\int_a^b h(t)_a D_t^{\beta_1(\cdot,\cdot)}\partial_3 L[x]_\gamma^{\alpha,\beta}(t)dt - \left[h(t)_a I_t^{1-\beta_1(\cdot,\cdot)}\partial_3 L[x]_\gamma^{\alpha,\beta}(t)\right]_{t=b}\right.$$

$$- \left(\int_T^b h(t)_T D_t^{\beta_1(\cdot,\cdot)}\partial_3 L[x]_\gamma^{\alpha,\beta}(t)dt - \left[h(t)_T I_t^{1-\beta_1(\cdot,\cdot)}\partial_3 L[x]_\gamma^{\alpha,\beta}(t)\right]_{t=b}\right.$$

$$\left.\left. + \left[h(t)_T I_t^{1-\beta_1(\cdot,\cdot)}\partial_3 L[x]_\gamma^{\alpha,\beta}(t)\right]_{t=T}\right)\right].$$

Unfolding these integrals, and considering the fractional operator $D_{\overline{\gamma^1}}^{\beta_1,\alpha_1}$ with $\overline{\gamma^1} = (\gamma_2^1, \gamma_1^1)$, then the previous term is equal to

$$\int_a^T h(t)D_{\overline{\gamma^1}}^{\beta_1(\cdot,\cdot),\alpha_1(\cdot,\cdot)}\partial_3 L[x]_\gamma^{\alpha,\beta}(t)dt$$

$$+ \int_T^b \gamma_2^1 h(t)\left[{}_a D_t^{\beta_1(\cdot,\cdot)}\partial_3 L[x]_\gamma^{\alpha,\beta}(t) - {}_T D_t^{\beta_1(\cdot,\cdot)}\partial_3 L[x]_\gamma^{\alpha,\beta}(t)\right]dt$$

$$+ h(T) \left[\gamma_1^1 \, {}_t I_T^{1-\alpha_1(\cdot,\cdot)} \partial_3 L[x]_\gamma^{\alpha,\beta}(t) - \gamma_2^1 \, {}_T I_t^{1-\beta_1(\cdot,\cdot)} \partial_3 L[x]_\gamma^{\alpha,\beta}(t) \right]_{t=T}$$

$$- h(b) \gamma_2^1 \left[{}_a I_t^{1-\beta_1(\cdot,\cdot)} \partial_3 L[x]_\gamma^{\alpha,\beta}(t) - {}_T I_t^{1-\beta_1(\cdot,\cdot)} \partial_3 L[x]_\gamma^{\alpha,\beta}(t) \right]_{t=b}.$$

Considering the third addend of the integral function (4.23), for $i = 2$, we get

$$\int_a^T \partial_4 L[x]_\gamma^{\alpha,\beta}(t) \, {}^C D_{\gamma^2}^{\alpha_2(\cdot,\cdot),\beta_2(\cdot,\cdot)} h(t)dt = \gamma_1^2 \int_a^T \partial_4 L[x]_\gamma^{\alpha,\beta}(t) \, {}_a^C D_t^{\alpha_2(\cdot,\cdot)} h(t)dt$$

$$+ \gamma_2^2 \left[\int_a^b \partial_4 L[x]_\gamma^{\alpha,\beta}(t) \, {}_t^C D_b^{\beta_2(\cdot,\cdot)} h(t)dt - \int_T^b \partial_4 L[x]_\gamma^{\alpha,\beta}(t) \, {}_t^C D_b^{\beta_2(\cdot,\cdot)} h(t)dt \right]$$

$$= \gamma_1^2 \left[\int_a^T h(t) \, {}_t D_T^{\alpha_2(\cdot,\cdot)} \partial_4 L[x]_\gamma^{\alpha,\beta}(t)dt \right.$$

$$+ \left[h^{(1)}(t) \, {}_t I_T^{2-\alpha_2(\cdot,\cdot)} \partial_4 L[x]_\gamma^{\alpha,\beta}(t) - h(t) \frac{d}{dt} \, {}_t I_T^{2-\alpha_2(\cdot,\cdot)} \partial_4 L[x]_\gamma^{\alpha,\beta}(t) \right]_{t=T} \right]$$

$$+ \gamma_2^2 \left[\int_a^b h(t) \, {}_a D_t^{\beta_2(\cdot,\cdot)} \partial_4 L[x]_\gamma^{\alpha,\beta}(t)dt \right.$$

$$+ \left[h^{(1)}(t) \, {}_a I_t^{2-\beta_2(\cdot,\cdot)} \partial_4 L[x]_\gamma^{\alpha,\beta}(t) - h(t) \frac{d}{dt} \, {}_a I_t^{2-\beta_2(\cdot,\cdot)} \partial_4 L[x]_\gamma^{\alpha,\beta}(t) \right]_{t=b}$$

$$- \int_T^b h(t) \, {}_T D_t^{\beta_2(\cdot,\cdot)} \partial_4 L[x]_\gamma^{\alpha,\beta}(t)dt$$

$$- \left[h^{(1)}(t) \, {}_T I_t^{2-\beta_2(\cdot,\cdot)} \partial_4 L[x]_\gamma^{\alpha,\beta}(t) - h(t) \frac{d}{dt} \, {}_T I_t^{2-\beta_2(\cdot,\cdot)} \partial_4 L[x]_\gamma^{\alpha,\beta}(t) \right]_{t=b}$$

$$+ \left[h^{(1)}(t) \, {}_T I_t^{2-\beta_2(\cdot,\cdot)} \partial_4 L[x]_\gamma^{\alpha,\beta}(t) - h(t) \frac{d}{dt} \, {}_T I_t^{2-\beta_2(\cdot,\cdot)} \partial_4 L[x]_\gamma^{\alpha,\beta}(t) \right]_{t=T} \right].$$

Again, with the auxiliary operator $D_{\overline{\gamma^2}}^{\beta_2,\alpha_2}$, with $\overline{\gamma^2} = (\gamma_2^2, \gamma_1^2)$, we obtain

$$\int_a^T h(t) D_{\overline{\gamma^2}}^{\beta_2(\cdot,\cdot),\alpha_2(\cdot,\cdot)} \partial_4 L[x]_\gamma^{\alpha,\beta}(t)dt$$

$$+ \int_T^b \gamma_2^2 h(t) \left[{}_a D_t^{\beta_2(\cdot,\cdot)} \partial_4 L[x]_\gamma^{\alpha,\beta}(t) - {}_T D_t^{\beta_2(\cdot,\cdot)} \partial_4 L[x]_\gamma^{\alpha,\beta}(t) \right] dt$$

$$+ \left[h^{(1)}(t) \left(\gamma_1^2 \, {}_t I_T^{2-\alpha_2(\cdot,\cdot)} \partial_4 L[x]_\gamma^{\alpha,\beta}(t) + \gamma_2^2 \, {}_T I_t^{2-\beta_2(\cdot,\cdot)} \partial_4 L[x]_\gamma^{\alpha,\beta}(t) \right) \right]_{t=T}$$

$$- \left[h(t) \left(\gamma_1^2 \frac{d}{dt} \, {}_t I_T^{2-\alpha_2(\cdot,\cdot)} \partial_4 L[x]_\gamma^{\alpha,\beta}(t) + \gamma_2^2 \frac{d}{dt} \, {}_T I_t^{2-\beta_2(\cdot,\cdot)} \partial_4 L[x]_\gamma^{\alpha,\beta}(t) \right) \right]_{t=T}$$

$$+ \left[h^{(1)}(t) \gamma_2^2 \left({}_a I_t^{2-\beta_2(\cdot,\cdot)} \partial_4 L[x]_\gamma^{\alpha,\beta}(t) - {}_T I_t^{2-\beta_2(\cdot,\cdot)} \partial_4 L[x]_\gamma^{\alpha,\beta}(t) \right) \right]_{t=b}$$

$$- \left[h(t) \gamma_2^2 \left(\frac{d}{dt} \, {}_a I_t^{2-\beta_2(\cdot,\cdot)} \partial_4 L[x]_\gamma^{\alpha,\beta}(t) - \frac{d}{dt} \, {}_T I_t^{2-\beta_2(\cdot,\cdot)} \partial_4 L[x]_\gamma^{\alpha,\beta}(t) \right) \right]_{t=b}.$$

Now, consider the general case

$$\int_a^T \partial_{i+2} L[x]_\gamma^{\alpha,\beta}(t)\, ^C D_{\gamma^i}^{\alpha_i(\cdot,\cdot),\beta_i(\cdot,\cdot)} h(t)dt,$$

$i = 3, \ldots, n$. Then, we obtain

$$\gamma_1^i \left[\int_a^T h(t)\, _t D_T^{\alpha_i(\cdot,\cdot)} \partial_{i+2} L[x]_\gamma^{\alpha,\beta}(t)dt \right.$$

$$\left. + \left[\sum_{k=0}^{i-1} (-1)^k h^{(i-1-k)}(t) \frac{d^k}{dt^k}\, _t I_T^{i-\alpha_i(\cdot,\cdot)} \partial_{i+2} L[x]_\gamma^{\alpha,\beta}(t)t) \right]_{t=T} \right]$$

$$+ \gamma_2^i \left[\int_a^b h(t)\, _a D_t^{\beta_i(\cdot,\cdot)} \partial_{i+2} L[x]_\gamma^{\alpha,\beta}(t)dt \right.$$

$$+ \left[\sum_{k=0}^{i-1} (-1)^{i+k} h^{(i-1-k)}(t) \frac{d^k}{dt^k}\, _a I_t^{i-\beta_i(\cdot,\cdot)} \partial_{i+2} L[x]_\gamma^{\alpha,\beta}(t) \right]_{t=b}$$

$$- \int_T^b h(t)\, _T D_t^{\beta_i(\cdot,\cdot)} \partial_{i+2} L[x]_\gamma^{\alpha,\beta}(t)dt$$

$$\left. - \left[\sum_{k=0}^{i-1} (-1)^{i+k} h^{(i-1-k)}(t) \frac{d^k}{dt^k}\, _T I_t^{i-\beta_i(\cdot,\cdot)} \partial_{i+2} L[x]_\gamma^{\alpha,\beta}(t) \right]_{t=T}^{t=b} \right].$$

Unfolding these integrals, we obtain

$$\int_a^T h(t) D_{\gamma^i}^{\beta_i(\cdot,\cdot),\alpha_i(\cdot,\cdot)} \partial_{i+2} L[x]_\gamma^{\alpha,\beta}(t)dt$$

$$+ \int_T^b \gamma_2^i h(t) \left[_a D_t^{\beta_i(\cdot,\cdot)} \partial_{i+2} L[x]_\gamma^{\alpha,\beta}(t) - _T D_t^{\beta_i(\cdot,\cdot)} \partial_{i+2} L[x]_\gamma^{\alpha,\beta}(t) \right] dt$$

$$+ h^{(i-1)}(T) \left[\gamma_1^i\, _t I_T^{i-\alpha_i(\cdot,\cdot)} \partial_{i+2} L[x]_\gamma^{\alpha,\beta}(t) + \gamma_2^i (-1)^i\, _T I_t^{i-\beta_i(\cdot,\cdot)} \partial_{i+2} L[x]_\gamma^{\alpha,\beta}(t) \right]_{t=T}$$

$$+ h^{(i-1)}(b) \gamma_2^i (-1)^i \left[_a I_t^{i-\beta_i(\cdot,\cdot)} \partial_{i+2} L[x]_\gamma^{\alpha,\beta}(t) - _T I_t^{i-\beta_i(\cdot,\cdot)} \partial_{i+2} L[x]_\gamma^{\alpha,\beta}(t) \right]_{t=b}$$

$$+ h^{(i-2)}(T) \left[\gamma_1^i (-1)^1 \frac{d}{dt}\, _t I_T^{i-\alpha_i(\cdot,\cdot)} \partial_{i+2} L[x]_\gamma^{\alpha,\beta}(t) \right.$$

$$\left. + \gamma_2^i (-1)^{i+1} \frac{d}{dt}\, _T I_t^{i-\beta_i(\cdot,\cdot)} \partial_{i+2} L[x]_\gamma^{\alpha,\beta}(t) \right]_{t=T}$$

$$+ h^{(i-2)}(b) \gamma_2^i (-1)^{i+1} \left[\frac{d}{dt}\, _a I_t^{i-\beta_i(\cdot,\cdot)} \partial_{i+2} L[x]_\gamma^{\alpha,\beta}(t) \right.$$

$$\left. - \frac{d}{dt}\, _T I_t^{i-\beta_i(\cdot,\cdot)} \partial_{i+2} L[x]_\gamma^{\alpha,\beta}(t) \right]_{t=b}$$

$$+ \ldots + h(T) \left[\gamma_1^i (-1)^{i-1} \frac{d^{i-1}}{dt^{i-1}} {}_t I_T^{i-\alpha_i(\cdot,\cdot)} \partial_{i+2} L[x]_\gamma^{\alpha,\beta}(t) \right.$$

$$\left. + \gamma_2^i (-1)^{2i-1} \frac{d^{i-1}}{dt^{i-1}} {}_T I_t^{i-\beta_i(\cdot,\cdot)} \partial_{i+2} L[x]_\gamma^{\alpha,\beta}(t) \right]_{t=T}$$

$$+ h(b) \gamma_2^i (-1)^{2i-1} \left[\frac{d^{i-1}}{dt^{i-1}} {}_a I_t^{i-\beta_i(\cdot,\cdot)} \partial_{i+2} L[x]_\gamma^{\alpha,\beta}(t) \right.$$

$$\left. - \frac{d^{i-1}}{dt^{i-1}} {}_T I_t^{i-\beta_i(\cdot,\cdot)} \partial_{i+2} L[x]_\gamma^{\alpha,\beta}(t) \right]_{t=b}.$$

Substituting all the relations into Eq. (4.23), we obtain that

$$0 = \int_a^T h(t) \left(\partial_2 L[x]_\gamma^{\alpha,\beta}(t) + \sum_{i=1}^n D_{\gamma^i}^{\beta_i(\cdot,\cdot),\alpha_i(\cdot,\cdot)} \partial_{i+2} L[x]_\gamma^{\alpha,\beta}(t) \right) dt$$

$$+ \int_T^b h(t) \sum_{i=1}^n \gamma_2^i \left[{}_a D_t^{\beta_i(\cdot,\cdot)} \partial_{i+2} L[x]_\gamma^{\alpha,\beta}(t) - {}_T D_t^{\beta_i(\cdot,\cdot)} \partial_{i+2} L[x]_\gamma^{\alpha,\beta}(t) \right] dt$$

$$+ \sum_{j=0}^{n-1} h^{(j)}(T) \sum_{i=j+1}^n \left[\gamma_1^i (-1)^{i-1-j} \frac{d^{i-1-j}}{dt^{i-1-j}} {}_t I_T^{i-\alpha_i(\cdot,\cdot)} \partial_{i+2} L[x]_\gamma^{\alpha,\beta}(t) \right.$$

$$\left. + \gamma_2^i (-1)^{j+1} \frac{d^{i-1-j}}{dt^{i-1-j}} {}_T I_t^{i-\beta_i(\cdot,\cdot)} \partial_{i+2} L[x]_\gamma^{\alpha,\beta}(t) \right]_{t=T}$$

$$+ \sum_{j=0}^{n-1} h^{(j)}(b) \sum_{i=j+1}^n \gamma_2^i (-1)^{j+1} \left[\frac{d^{i-1-j}}{dt^{i-1-j}} {}_a I_t^{i-\beta_i(\cdot,\cdot)} \partial_{i+2} L[x]_\gamma^{\alpha,\beta}(t) \right.$$

$$\left. - \frac{d^{i-1-j}}{dt^{i-1-j}} {}_T I_t^{i-\beta_i(\cdot,\cdot)} \partial_{i+2} L[x]_\gamma^{\alpha,\beta}(t) \right]_{t=b} + h(T) \partial_2 \phi(T, x(T)) \tag{4.24}$$

$$+ \Delta T \left[L[x]_\gamma^{\alpha,\beta}(T) + \partial_1 \phi(T, x(T)) + \partial_2 \phi(T, x(T)) x'(T) \right].$$

We obtain the fractional Euler–Lagrange equations (4.20)–(4.21) and the transversality conditions (4.22) applying the fundamental lemma of the calculus of variations (see, e.g., van Brunt [25]) for appropriate choices of variations.

Remark 48 When $n = 1$, functional (4.18) takes the form

$$\mathcal{J}(x, T) = \int_a^T L\left(t, x(t), {}^C D_\gamma^{\alpha(\cdot,\cdot),\beta(\cdot,\cdot)} x(t) \right) dt + \phi(T, x(T)),$$

and the fractional Euler–Lagrange equations (4.20)–(4.21) coincide with those of Theorem 45.

Considering the increment ΔT on time T, and the consequent increment Δx_T on x, given by

$$\Delta x_T = (x + h)(T + \Delta T) - x(T), \tag{4.25}$$

in the next theorem, we rewrite the transversality conditions (4.22) in terms of these increments.

Theorem 49 *If* (x, T) *minimizes functional* $\mathcal{J}$ *defined by* (4.19), *then* (x, T) *satisfies the Euler–Lagrange equations* (4.20) *and* (4.21) *and the following transversality conditions:*

$$
\begin{cases}
\partial_1 \phi(T, x(T)) - x'(T) \sum_{i=1}^{n} \left[\gamma_1^i (-1)^{i-1} \dfrac{d^{i-1}}{dt^{i-1}} {}_t I_T^{i-\alpha_i(\cdot,\cdot)} \partial_{i+2} L[x]_\gamma^{\alpha,\beta}(t) \right. \\
\qquad \left. - \gamma_2^i \dfrac{d^{i-1}}{dt^{i-1}} T I_t^{i-\beta_i(\cdot,\cdot)} \partial_{i+2} L[x]_\gamma^{\alpha,\beta}(t) + L[x]_\gamma^{\alpha,\beta}(T) \right]_{t=T} = 0, \\[2ex]
\displaystyle\sum_{i=1}^{n} \left[\gamma_1^i (-1)^{i-1} \dfrac{d^{i-1}}{dt^{i-1}} {}_t I_T^{i-\alpha_i(\cdot,\cdot)} \partial_{i+2} L[x]_\gamma^{\alpha,\beta}(t) \right. \\
\qquad \left. - \gamma_2^i \dfrac{d^{i-1}}{dt^{i-1}} T I_t^{i-\beta_i(\cdot,\cdot)} \partial_{i+2} L[x]_\gamma^{\alpha,\beta}(t) \right]_{t=T} + \partial_2 \phi(T, x(T)) = 0, \\[2ex]
\displaystyle\sum_{i=j+1}^{n} \left[\gamma_1^i (-1)^{i-1-j} \dfrac{d^{i-1-j}}{dt^{i-1-j}} {}_t I_T^{i-\alpha_i(\cdot,\cdot)} \partial_{i+2} L[x]_\gamma^{\alpha,\beta}(t) \right. \\
\qquad \left. + \gamma_2^i (-1)^{j+1} \dfrac{d^{i-1-j}}{dt^{i-1-j}} T I_t^{i-\beta_i(\cdot,\cdot)} \partial_{i+2} L[x]_\gamma^{\alpha,\beta}(t) \right]_{t=T} = 0, \quad \forall j = 2, \ldots, n-1, \\[2ex]
\displaystyle\sum_{i=j+1}^{n} \left[\gamma_2^i (-1)^{j+1} \left[\dfrac{d^{i-1-j}}{dt^{i-1-j}} {}_a I_t^{i-\beta_i(\cdot,\cdot)} \partial_{i+2} L[x]_\gamma^{\alpha,\beta}(t) \right. \right. \\
\qquad \left. \left. - \dfrac{d^{i-1-j}}{dt^{i-1-j}} T I_t^{i-\beta_i(\cdot,\cdot)} \partial_{i+2} L[x]_\gamma^{\alpha,\beta}(t) \right] \right]_{t=b} = 0, \quad \forall j = 0, \ldots, n-1.
\end{cases}
$$

Proof Using Taylor's expansion up to first order, and restricting the set of variations to those for which $h'(T) = 0$, we obtain that

$$(x + h)(T + \Delta T) = (x + h)(T) + x'(T)\Delta T + O(\Delta T)^2.$$

According to the increment on x given by (4.25), we get

$$h(T) = \Delta x_T - x'(T)\Delta T + O(\Delta T)^2.$$

From substitution of this expression into (4.24), and by using appropriate choices of variations, we obtain the intended transversality conditions.

4.3.2 Example

We provide an illustrative example. It is covered by Theorem 47.

Let $p_{n-1}(t)$ be a polynomial of degree $n - 1$. If $\alpha, \beta : [0, b]^2 \to (n - 1, n)$ are the fractional orders, then ${}^C D_\gamma^{\alpha(\cdot,\cdot),\beta(\cdot,\cdot)} p_{n-1}(t) = 0$ since $p_{n-1}^{(n)}(t) = 0$ for all t. Consider the functional

$$J(x, T) = \int_0^T \left[\left({}^C D_\gamma^{\alpha(\cdot,\cdot),\beta(\cdot,\cdot)} x(t) \right)^2 + (x(t) - p_{n-1}(t))^2 - t - 1 \right] dt + T^2$$

subject to the initial constraints

$$x(0) = p_{n-1}(0) \quad \text{and} \quad x^{(k)}(0) = p_{n-1}^{(k)}(0), \ k = 1, \ldots, n - 1.$$

Observe that, for all $t \in [0, b]$,

$$\partial_i L[p_{n-1}]_\gamma^{\alpha,\beta}(t) = 0, \quad i = 2, 3.$$

Thus, function $x \equiv p_{n-1}$ and the final time $T = 1$ satisfy the necessary optimality conditions of Theorem 47. We also remark that, for any curve x, one has

$$J(x, T) > \int_0^T [-t - 1] \, dt + T^2 = \frac{T^2}{2} - T,$$

which attains a minimum value $-1/2$ at $T = 1$. Since $J(p_{n-1}, 1) = -1/2$, we conclude that $(p_{n-1}, 1)$ is the (global) minimizer of J.

4.4 Variational Problems with Time Delay

In this section, we consider fractional variational problems with time delay. As mentioned in Machado [15], "we verify that a fractional derivative requires an infinite number of samples capturing, therefore, all the signal history, contrary to what happens with integer-order derivatives that are merely local operators. This fact motivates the evaluation of calculation strategies based on delayed signal samples." This subject has already been studied for constant fractional order [1, 3, 12]. However, for a variable fractional order, it is, to the authors' best knowledge, an open question. We also refer to the works [5, 6, 13, 26], where fractional differential equations are considered with a time delay.

In Sect. 4.4.1, we deduce necessary optimality conditions when the Lagrangian depends on a time delay (Theorem 50). One illustrative example is discussed in Sect. 4.4.2.

4.4.1 Necessary Optimality Conditions

For simplicity of presentation, we consider fractional orders $\alpha, \beta : [a, b]^2 \to (0, 1)$. Using similar arguments, the problem can be easily generalized for higher-order derivatives. Let $\sigma > 0$ and define the vector

$$\sigma[x]_\gamma^{\alpha,\beta}(t) = \left(t, x(t), {}^C D_\gamma^{\alpha(\cdot,\cdot),\beta(\cdot,\cdot)} x(t), x(t-\sigma)\right).$$

For the domain of the functional, we consider the set

$$D_\sigma = \left\{(x,t) \in C^1([a-\sigma,b]) \times [a,b] : {}^C D_\gamma^{\alpha(\cdot,\cdot),\beta(\cdot,\cdot)} x \in C([a,b])\right\}.$$

Let $\mathcal{J} : D_\sigma \to \mathbb{R}$ be the functional defined by

$$\mathcal{J}(x,T) = \int_a^T L_\sigma[x]_\gamma^{\alpha,\beta}(t) + \phi(T, x(T)), \tag{4.26}$$

where we assume again that the Lagrangian L and the payoff term ϕ are differentiable. The optimization problem with time delay is the following:

Problem 3 Minimize functional (4.26) on D_σ subject to the boundary condition

$$x(t) = \varphi(t)$$

for all $t \in [a-\sigma, a]$, where φ is a given (fixed) function.

We now state and prove the Euler–Lagrange equations for this problem.

Theorem 50 *Suppose that (x,T) gives a local minimum to functional (4.26) on D_σ. If $\sigma \geq T-a$, then (x,T) satisfies*

$$\partial_2 L_\sigma[x]_\gamma^{\alpha,\beta}(t) + D_{\overline{\gamma}}^{\beta(\cdot,\cdot),\alpha(\cdot,\cdot)} \partial_3 L_\sigma[x]_\gamma^{\alpha,\beta}(t) = 0, \tag{4.27}$$

for $t \in [a,T]$ and

$$\gamma_2 \left({}_a D_t^{\beta(\cdot,\cdot)} \partial_3 L_\sigma[x]_\gamma^{\alpha,\beta}(t) - {}_T D_t^{\beta(\cdot,\cdot)} \partial_3 L_\sigma[x]_\gamma^{\alpha,\beta}(t) \right) = 0, \tag{4.28}$$

for $t \in [T,b]$. Moreover, (x,T) satisfies

$$\begin{cases} L_\sigma[x]_\gamma^{\alpha,\beta}(T) + \partial_1 \phi(T, x(T)) + \partial_2 \phi(T, x(T)) x'(T) = 0, \\ \left[\gamma_1 \, {}_t I_T^{1-\alpha(\cdot,\cdot)} \partial_3 L_\sigma[x]_\gamma^{\alpha,\beta}(t) - \gamma_2 \, {}_T I_t^{1-\beta(\cdot,\cdot)} \partial_3 L_\sigma[x]_\gamma^{\alpha,\beta}(t) \right]_{t=T} \\ \qquad + \partial_2 \phi(T, x(T)) = 0, \\ \gamma_2 \left[{}_T I_t^{1-\beta(\cdot,\cdot)} \partial_3 L_\sigma[x]_\gamma^{\alpha,\beta}(t) - {}_a I_t^{1-\beta(\cdot,\cdot)} \partial_3 L_\sigma[x]_\gamma^{\alpha,\beta}(t) \right]_{t=b} = 0. \end{cases} \tag{4.29}$$

If $\sigma < T-a$, then Eq. (4.27) is replaced by the two following ones:

$$\partial_2 L_\sigma[x]_\gamma^{\alpha,\beta}(t) + D_{\overline{\gamma}}^{\beta(\cdot,\cdot),\alpha(\cdot,\cdot)} \partial_3 L_\sigma[x]_\gamma^{\alpha,\beta}(t) + \partial_4 L_\sigma[x]_\gamma^{\alpha,\beta}(t+\sigma) = 0, \tag{4.30}$$

for $t \in [a, T - \sigma]$, and

$$\partial_2 L_\sigma[x]^{\alpha,\beta}_\gamma(t) + D^{\beta(\cdot,\cdot),\alpha(\cdot,\cdot)}_{\overline{\gamma}}\partial_3 L_\sigma[x]^{\alpha,\beta}_\gamma(t) = 0, \qquad (4.31)$$

for $t \in [T - \sigma, T]$.

Proof Consider variations of the solution $(x + \epsilon h, T + \epsilon \Delta T)$, where $h \in C^1([a - \sigma, b]; \mathbb{R})$ is such that $h(t) = 0$ for all $t \in [a - \sigma, a]$, and ϵ, ΔT are two reals. If we define $j(\epsilon) = \mathcal{J}(x + \epsilon h, T + \epsilon \Delta T)$, then $j'(0) = 0$, that is,

$$\int_a^T \left(\partial_2 L_\sigma[x]^{\alpha,\beta}_\gamma(t)h(t) + \partial_3 L_\sigma[x]^{\alpha,\beta}_\gamma(t)\,{}^C D^{\alpha(\cdot,\cdot),\beta(\cdot,\cdot)}_\gamma h(t) \right.$$

$$\left. + \partial_4 L_\sigma[x]^{\alpha,\beta}_\gamma(t)h(t - \sigma) \right)dt + L_\sigma[x]^{\alpha,\beta}_\gamma(T)\Delta T$$

$$+ \partial_1\phi(T, x(T))\,\Delta T + \partial_2\phi(T, x(T))\left[h(T) + x'(T)\Delta T\right] = 0. \quad (4.32)$$

First, suppose that $\sigma \geq T - a$. In this case, since

$$\int_a^T \partial_4 L_\sigma[x]^{\alpha,\beta}_\gamma(t)h(t - \sigma)\,dt = \int_{a-\sigma}^{T-\sigma} \partial_4 L_\sigma[x]^{\alpha,\beta}_\gamma(t + \sigma)h(t)\,dt$$

and $h \equiv 0$ on $[a - \sigma, a]$, this term vanishes in (4.32) and we obtain Eq. (4.12). The rest of the proof is similar to the one presented in Sect. 4.2, Theorem 45, and we obtain (4.27)–(4.29). Suppose now that $\sigma < T - a$. In this case, we have that

$$\int_a^T \partial_4 L_\sigma[x]^{\alpha,\beta}_\gamma(t)h(t - \sigma)\,dt = \int_{a-\sigma}^{T-\sigma} \partial_4 L_\sigma[x]^{\alpha,\beta}_\gamma(t + \sigma)h(t)\,dt$$

$$= \int_a^{T-\sigma} \partial_4 L_\sigma[x]^{\alpha,\beta}_\gamma(t + \sigma)h(t)\,dt.$$

Next, we evaluate the integral

$$\int_a^T \partial_3 L_\sigma[x]^{\alpha,\beta}_\gamma(t)\,{}^C D^{\alpha(\cdot,\cdot),\beta(\cdot,\cdot)}_\gamma h(t)\,dt$$

$$= \int_a^{T-\sigma} \partial_3 L_\sigma[x]^{\alpha,\beta}_\gamma(t)\,{}^C D^{\alpha(\cdot,\cdot),\beta(\cdot,\cdot)}_\gamma h(t)\,dt$$

$$+ \int_{T-\sigma}^T \partial_3 L_\sigma[x]^{\alpha,\beta}_\gamma(t)\,{}^C D^{\alpha(\cdot,\cdot),\beta(\cdot,\cdot)}_\gamma h(t)\,dt.$$

For the first integral, integrating by parts, we have

$$\int_a^{T-\sigma} \partial_3 L_\sigma[x]_\gamma^{\alpha,\beta}(t)\,^C D_\gamma^{\alpha(\cdot,\cdot),\beta(\cdot,\cdot)} h(t)\,dt = \gamma_1 \int_a^{T-\sigma} \partial_3 L_\sigma[x]_\gamma^{\alpha,\beta}(t)\,_a^C D_t^{\alpha(\cdot,\cdot)} h(t)\,dt$$

$$+ \gamma_2 \left[\int_a^b \partial_3 L_\sigma[x]_\gamma^{\alpha,\beta}(t)\,_t^C D_b^{\beta(\cdot,\cdot)} h(t)\,dt - \int_{T-\sigma}^b \partial_3 L_\sigma[x]_\gamma^{\alpha,\beta}(t)\,_t^C D_b^{\beta(\cdot,\cdot)} h(t)\,dt \right]$$

$$= \int_a^{T-\sigma} h(t) \left[\gamma_{1t} D_{T-\sigma}^{\alpha(\cdot,\cdot)} \partial_3 L_\sigma[x]_\gamma^{\alpha,\beta}(t) + \gamma_{2a} D_t^{\beta(\cdot,\cdot)} \partial_3 L_\sigma[x]_\gamma^{\alpha,\beta}(t) \right] dt$$

$$+ \int_{T-\sigma}^b \gamma_2 h(t) \left[{}_a D_t^{\beta(\cdot,\cdot)} \partial_3 L_\sigma[x]_\gamma^{\alpha,\beta}(t) - {}_{T-\sigma} D_t^{\beta(\cdot,\cdot)} \partial_3 L_\sigma[x]_\gamma^{\alpha,\beta}(t) \right] dt$$

$$+ \left[h(t) \left[\gamma_{1t} I_{T-\sigma}^{1-\alpha(\cdot,\cdot)} \partial_3 L_\sigma[x]_\gamma^{\alpha,\beta}(t) - \gamma_{2T-\sigma} I_t^{1-\beta(\cdot,\cdot)} \partial_3 L_\sigma[x]_\gamma^{\alpha,\beta}(t) \right] \right]_{t=T-\sigma}$$

$$+ \left[\gamma_2 h(t) \left[-{}_a I_t^{1-\beta(\cdot,\cdot)} \partial_3 L_\sigma[x]_\gamma^{\alpha,\beta}(t) + {}_{T-\sigma} I_t^{1-\beta(\cdot,\cdot)} \partial_3 L_\sigma[x]_\gamma^{\alpha,\beta}(t) \right] \right]_{t=b}.$$

For the second integral, in a similar way, we deduce that

$$\int_{T-\sigma}^T \partial_3 L_\sigma[x]_\gamma^{\alpha,\beta}(t)\,^C D_\gamma^{\alpha(\cdot,\cdot),\beta(\cdot,\cdot)} h(t)\,dt$$

$$= \gamma_1 \left[\int_a^T \partial_3 L_\sigma[x]_\gamma^{\alpha,\beta}(t)\,_a^C D_t^{\alpha(\cdot,\cdot)} h(t)\,dt - \int_a^{T-\sigma} \partial_3 L_\sigma[x]_\gamma^{\alpha,\beta}(t)\,_a^C D_t^{\alpha(\cdot,\cdot)} h(t)\,dt \right]$$

$$+ \gamma_2 \left[\int_{T-\sigma}^b \partial_3 L_\sigma[x]_\gamma^{\alpha,\beta}(t)\,_t^C D_b^{\beta(\cdot,\cdot)} h(t)\,dt - \int_T^b \partial_3 L_\sigma[x]_\gamma^{\alpha,\beta}(t)\,_t^C D_b^{\beta(\cdot,\cdot)} h(t)\,dt \right]$$

$$= \int_a^{T-\sigma} \gamma_1 h(t) \left[{}_t D_T^{\alpha(\cdot,\cdot)} \partial_3 L_\sigma[x]_\gamma^{\alpha,\beta}(t) - {}_t D_{T-\sigma}^{\alpha(\cdot,\cdot)} \partial_3 L_\sigma[x]_\gamma^{\alpha,\beta}(t) \right] dt$$

$$+ \int_{T-\sigma}^T h(t) \left[\gamma_{1t} D_T^{\alpha(\cdot,\cdot)} \partial_3 L_\sigma[x]_\gamma^{\alpha,\beta}(t) + \gamma_{2T-\sigma} D_t^{\beta(\cdot,\cdot)} \partial_3 L_\sigma[x]_\gamma^{\alpha,\beta}(t) \right] dt$$

$$+ \int_T^b \gamma_2 h(t) \left[{}_{T-\sigma} D_t^{\beta(\cdot,\cdot)} \partial_3 L_\sigma[x]_\gamma^{\alpha,\beta}(t) - {}_T D_t^{\beta(\cdot,\cdot)} \partial_3 L_\sigma[x]_\gamma^{\alpha,\beta}(t) \right] dt$$

$$+ \left[h(t) \left[-\gamma_{1t} I_{T-\sigma}^{1-\alpha(\cdot,\cdot)} \partial_3 L_\sigma[x]_\gamma^{\alpha,\beta}(t) + \gamma_{2T-\sigma} I_t^{1-\beta(\cdot,\cdot)} \partial_3 L_\sigma[x]_\gamma^{\alpha,\beta}(t) \right] \right]_{t=T-\sigma}$$

$$+ \left[h(t) \left[\gamma_{1t} I_T^{1-\alpha(\cdot,\cdot)} \partial_3 L_\sigma[x]_\gamma^{\alpha,\beta}(t) - \gamma_{2T} I_t^{1-\beta(\cdot,\cdot)} \partial_3 L_\sigma[x]_\gamma^{\alpha,\beta}(t) \right] \right]_{t=T}$$

$$+ \left[\gamma_2 h(t) \left[-{}_{T-\sigma} I_t^{1-\beta(\cdot,\cdot)} \partial_3 L_\sigma[x]_\gamma^{\alpha,\beta}(t) + {}_T I_t^{1-\beta(\cdot,\cdot)} \partial_3 L_\sigma[x]_\gamma^{\alpha,\beta}(t) \right] \right]_{t=b}.$$

Replacing the above equalities into (4.32), we prove that

$$0 = \int_a^{T-\sigma} h(t) \left[\partial_2 L_\sigma[x]_\gamma^{\alpha,\beta}(t) + D_{\overline{\gamma}}^{\beta(\cdot,\cdot),\alpha(\cdot,\cdot)} \partial_3 L_\sigma[x]_\gamma^{\alpha,\beta}(t) + \partial_4 L_\sigma[x]_\gamma^{\alpha,\beta}(t+\sigma) \right] dt$$

$$+ \int_{T-\sigma}^T h(t) \left[\partial_2 L_\sigma[x]_\gamma^{\alpha,\beta}(t) + D_{\overline{\gamma}}^{\beta(\cdot,\cdot),\alpha(\cdot,\cdot)} \partial_3 L_\sigma[x]_\gamma^{\alpha,\beta}(t) \right] dt$$

$$+ \int_T^b \gamma_2 h(t) \left[{}_a D_t^{\beta(\cdot,\cdot)} \partial_3 L_\sigma[x]_\gamma^{\alpha,\beta}(t) - {}_T D_t^{\beta(\cdot,\cdot)} \partial_3 L_\sigma[x]_\gamma^{\alpha,\beta}(t) \right] dt$$

$$+ h(T) \left[\gamma_1 \, _t I_T^{1-\alpha(\cdot,\cdot)} \partial_3 L_\sigma[x]_\gamma^{\alpha,\beta}(t) - \gamma_2 \, _T I_t^{1-\beta(\cdot,\cdot)} \partial_3 L_\sigma[x]_\gamma^{\alpha,\beta}(t) + \partial_2 \phi(t, x(t)) \right]_{t=T}$$

$$+ \Delta T \left[L_\sigma[x]_\gamma^{\alpha,\beta}(t) + \partial_1 \phi(t, x(t)) + \partial_2 \phi(t, x(t)) x'(t) \right]_{t=T}$$

$$+ h(b) \left[\gamma_2 \left(_T I_t^{1-\beta(\cdot,\cdot)} \partial_3 L_\sigma[x]_\gamma^{\alpha,\beta}(t) - a I_t^{1-\beta(\cdot,\cdot)} \partial_3 L_\sigma[x]_\gamma^{\alpha,\beta}(t) \right) \right]_{t=b}.$$

By the arbitrariness of h in $[a, b]$ and of ΔT, we obtain Eqs. (4.28)–(4.31).

4.4.2 Example

Let $\alpha, \beta : [0, b]^2 \rightarrow (0, 1)$, $\sigma = 1$, f be a function of class C^1, and $\hat{f}(t) = {}^C D_\gamma^{\alpha(\cdot,\cdot),\beta(\cdot,\cdot)} f(t)$. Consider the following problem of the calculus of variations:

$$\mathcal{J}(x, T) = \int_0^T \left[\left({}^C D_\gamma^{\alpha(\cdot,\cdot),\beta(\cdot,\cdot)} x(t) - \hat{f}(t) \right)^2 + (x(t) - f(t))^2 \right.$$
$$\left. + (x(t-1) - f(t-1))^2 - t - 2 \right] dt + T^2 \rightarrow \min$$

subject to the condition $x(t) = f(t)$ for all $t \in [-1, 0]$.

In this case, we can easily verify that $(x, T) = (f, 2)$ satisfies all the conditions in Theorem 50 and that it is actually the (global) minimizer of the problem.

4.5 Isoperimetric Problems

Isoperimetric problems are optimization problems that consist in minimizing or maximizing a cost functional subject to an integral constraint. From the variational problem with dependence on a combined Caputo derivative of variable fractional order (see Definition 33) discussed in Sect. 4.2, here we study two variational problems subject to an additional integral constraint. In each of the problems, the terminal point in the cost integral, as well as the terminal state, is considered to be free, and we obtain corresponding natural boundary conditions.

In Sects. 4.5.1 and 4.5.2, we study necessary optimality conditions in order to determine the minimizers for each of the problems. We end this section with an example (Sect. 4.5.3).

For the two isoperimetric problems considered in the next sections, let D be the set given by (4.6) and $[x]_\gamma^{\alpha,\beta}(t)$ the vector (4.8).

Picking up the problem of the fractional calculus of variations with variable order, discussed in Sect. 4.2, we consider also a differentiable Lagrangian $L : [a, b] \times \mathbb{R}^2 \rightarrow \mathbb{R}$ and the functional $\mathcal{J} : D \rightarrow \mathbb{R}$ of the form

$$\mathcal{J}(x, T) = \int_a^T L[x]_\gamma^{\alpha,\beta}(t)dt + \phi(T, x(T)), \tag{4.33}$$

where the terminal cost function $\phi : [a, b] \times \mathbb{R} \to \mathbb{R}$ is of class C^1.

In the sequel, we need the auxiliary notation of the dual fractional derivative:

$$D_{\overline{\gamma},c}^{\beta(\cdot,\cdot),\alpha(\cdot,\cdot)} = \gamma_2 \, {}_aD_t^{\beta(\cdot,\cdot)} + \gamma_1 \, {}_tD_c^{\alpha(\cdot,\cdot)}, \quad \text{where} \quad \overline{\gamma} = (\gamma_2, \gamma_1) \quad \text{and} \quad c \in (a, b]. \tag{4.34}$$

With the functional $\mathcal{J}$, defined by (4.33), we consider two different isoperimetric problems.

4.5.1 Necessary Optimality Conditions I

The first fractional isoperimetric problem of the calculus of variations is Problem 4.

Problem 4 Determine the local minimizers of $\mathcal{J}$ over all $(x, T) \in D$ satisfying a boundary condition

$$x(a) = x_a \tag{4.35}$$

for a fixed $x_a \in \mathbb{R}$ and an integral constraint of the form

$$\int_a^T g[x]_\gamma^{\alpha,\beta}(t)dt = \psi(T), \tag{4.36}$$

where $g : C^1\left([a, b] \times \mathbb{R}^2\right) \to \mathbb{R}$ and $\psi : [a, b] \to \mathbb{R}$ are two differentiable functions. The terminal time T and terminal state $x(T)$ are free.

In this problem, the condition of the form (4.36) is called an isoperimetric constraint. The next theorem gives fractional necessary optimality conditions for Problem 4.

Theorem 51 *Suppose that (x, T) gives a local minimum for functional (4.33) on D subject to the boundary condition (4.35) and the isoperimetric constraint (4.36). If (x, T) does not satisfy the Euler–Lagrange equations with respect to the isoperimetric constraint, that is, if one of the two following conditions are not verified,*

$$\partial_2 g[x]_\gamma^{\alpha,\beta}(t) + D_{\overline{\gamma},T}^{\beta(\cdot,\cdot),\alpha(\cdot,\cdot)}\partial_3 g[x]_\gamma^{\alpha,\beta}(t) = 0, \quad t \in [a, T], \tag{4.37}$$

or

$$\gamma_2 \left[{}_aD_t^{\beta(\cdot,\cdot)}\partial_3 g[x]_\gamma^{\alpha,\beta}(t) - {}_TD_t^{\beta(\cdot,\cdot)}\partial_3 g[x]_\gamma^{\alpha,\beta}(t) \right] = 0, \quad t \in [T, b], \tag{4.38}$$

then there exists a constant λ such that if we define the function $F : [a, b] \times \mathbb{R}^2 \to \mathbb{R}$ by $F = L - \lambda g$, (x, T) satisfies the fractional Euler–Lagrange equations

$$\partial_2 F[x]_\gamma^{\alpha,\beta}(t) + D_{\bar\gamma,T}^{\beta(\cdot,\cdot),\alpha(\cdot,\cdot)} \partial_3 F[x]_\gamma^{\alpha,\beta}(t) = 0 \tag{4.39}$$

on the interval $[a, T]$ *and*

$$\gamma_2 \left({}_aD_t^{\beta(\cdot,\cdot)} \partial_3 F[x]_\gamma^{\alpha,\beta}(t) - {}_TD_t^{\beta(\cdot,\cdot)} \partial_3 F[x]_\gamma^{\alpha,\beta}(t) \right) = 0 \tag{4.40}$$

on the interval $[T, b]$. *Moreover,* (x, T) *satisfies the transversality conditions*

$$\begin{cases} F[x]_\gamma^{\alpha,\beta}(T) + \partial_1 \phi(T, x(T)) + \partial_2 \phi(T, x(T)) x'(T) + \lambda \psi'(T) = 0, \\ \left[\gamma_1 {}_tI_T^{1-\alpha(\cdot,\cdot)} \partial_3 F[x]_\gamma^{\alpha,\beta}(t) - \gamma_2 {}_TI_t^{1-\beta(\cdot,\cdot)} \partial_3 F[x]_\gamma^{\alpha,\beta}(t) \right]_{t=T} + \partial_2 \phi(T, x(T)) = 0, \\ \gamma_2 \left[{}_TI_t^{1-\beta(\cdot,\cdot)} \partial_3 F[x]_\gamma^{\alpha,\beta}(t) - {}_aI_t^{1-\beta(\cdot,\cdot)} \partial_3 F[x]_\gamma^{\alpha,\beta}(t) \right]_{t=b} = 0. \end{cases}$$
$$\tag{4.41}$$

Proof Consider variations of the optimal solution (x, T) of the type

$$(x^*, T^*) = (x + \epsilon_1 h_1 + \epsilon_2 h_2, T + \epsilon_1 \Delta T), \tag{4.42}$$

where, for each $i \in \{1, 2\}$, $\epsilon_i \in \mathbb{R}$ is a small parameter, $h_i \in C^1([a, b]; \mathbb{R})$ satisfies $h_i(a) = 0$, and $\Delta T \in \mathbb{R}$. The additional term $\epsilon_2 h_2$ must be selected so that the admissible variations (x^*, T^*) satisfy the isoperimetric constraint (4.36). For a fixed choice of h_i, let

$$i(\epsilon_1, \epsilon_2) = \int_a^{T+\epsilon_1 \Delta T} g[x^*]_\gamma^{\alpha,\beta}(t) dt - \psi(T + \epsilon_1 \Delta T).$$

For $\epsilon_1 = \epsilon_2 = 0$, we obtain that

$$i(0, 0) = \int_a^T g[x]_\gamma^{\alpha,\beta}(t) dt - \psi(T) = \psi(T) - \psi(T) = 0.$$

The derivative $\dfrac{\partial i}{\partial \epsilon_2}$ is given by

$$\frac{\partial i}{\partial \epsilon_2} = \int_a^{T+\epsilon_1 \Delta T} \left(\partial_2 g[x^*]_\gamma^{\alpha,\beta}(t) h_2(t) + \partial_3 g[x^*]_\gamma^{\alpha,\beta}(t) {}^C D_\gamma^{\alpha(\cdot,\cdot),\beta(\cdot,\cdot)} h_2(t) \right) dt.$$

For $\epsilon_1 = \epsilon_2 = 0$, one has

$$\frac{\partial i}{\partial \epsilon_2}\bigg|_{(0,0)} = \int_a^T \left(\partial_2 g[x]_\gamma^{\alpha,\beta}(t) h_2(t) + \partial_3 g[x]_\gamma^{\alpha,\beta}(t) {}^C D_\gamma^{\alpha(\cdot,\cdot),\beta(\cdot,\cdot)} h_2(t) \right) dt. \tag{4.43}$$

The second term in (4.43) can be written as

$$\int_a^T \partial_3 g[x]_\gamma^{\alpha,\beta}(t)^C D_\gamma^{\alpha(\cdot,\cdot),\beta(\cdot,\cdot)} h_2(t) dt$$

$$= \int_a^T \partial_3 g[x]_\gamma^{\alpha,\beta}(t) \left[\gamma_1 \, {}_a^C D_t^{\alpha(\cdot,\cdot)} h_2(t) + \gamma_2 \, {}_t^C D_b^{\beta(\cdot,\cdot)} h_2(t) \right] dt$$

$$= \gamma_1 \int_a^T \partial_3 g[x]_\gamma^{\alpha,\beta}(t) {}_a^C D_t^{\alpha(\cdot,\cdot)} h_2(t) dt$$

$$+ \gamma_2 \left[\int_a^b \partial_3 g[x]_\gamma^{\alpha,\beta}(t) {}_t^C D_b^{\beta(\cdot,\cdot)} h_2(t) dt - \int_T^b \partial_3 g[x]_\gamma^{\alpha,\beta}(t) {}_t^C D_b^{\beta(\cdot,\cdot)} h_2(t) dt \right].$$

$$(4.44)$$

Using the fractional integrating by parts formula, (4.44) is equal to

$$\int_a^T h_2(t) \left[\gamma_{1t} D_T^{\alpha(\cdot,\cdot)} \partial_3 g[x]_\gamma^{\alpha,\beta}(t) + \gamma_{2a} D_t^{\beta(\cdot,\cdot)} \partial_3 g[x]_\gamma^{\alpha,\beta}(t) \right] dt$$

$$+ \int_T^b \gamma_2 h_2(t) \left[{}_a D_t^{\beta(\cdot,\cdot)} \partial_3 g[x]_\gamma^{\alpha,\beta}(t) - {}_T D_t^{\beta(\cdot,\cdot)} \partial_3 g[x]_\gamma^{\alpha,\beta}(t) \right] dt$$

$$+ \left[h_2(t) \left(\gamma_{1t} I_T^{1-\alpha(\cdot,\cdot)} \partial_3 g[x]_\gamma^{\alpha,\beta}(t) - \gamma_{2T} I_t^{1-\beta(\cdot,\cdot)} \partial_3 g[x]_\gamma^{\alpha,\beta}(t) \right) \right]_{t=T}$$

$$+ \left[\gamma_2 h_2(t) \left({}_T I_t^{1-\beta(\cdot,\cdot)} \partial_3 g[x]_\gamma^{\alpha,\beta}(t) - {}_a I_t^{1-\beta(\cdot,\cdot)} \partial_3 g[x]_\gamma^{\alpha,\beta}(t) \right) \right]_{t=b}.$$

Substituting these relations into (4.43), and considering the fractional operator $D_{\overline{\gamma},c}^{\beta(\cdot,\cdot),\alpha(\cdot,\cdot)}$ as defined in (4.34), we obtain that

$$\left. \frac{\partial i}{\partial \epsilon_2} \right|_{(0,0)} = \int_a^T h_2(t) \left[\partial_2 g[x]_\gamma^{\alpha,\beta}(t) + D_{\overline{\gamma},T}^{\beta(\cdot,\cdot),\alpha(\cdot,\cdot)} \partial_3 g[x]_\gamma^{\alpha,\beta}(t) \right] dt$$

$$+ \int_T^b \gamma_2 h_2(t) \left[{}_a D_t^{\beta(\cdot,\cdot)} \partial_3 g[x]_\gamma^{\alpha,\beta}(t) - {}_T D_t^{\beta(\cdot,\cdot)} \partial_3 g[x]_\gamma^{\alpha,\beta}(t) \right] dt$$

$$+ \left[h_2(t) \left(\gamma_{1t} I_T^{1-\alpha(\cdot,\cdot)} \partial_3 g[x]_\gamma^{\alpha,\beta}(t) - \gamma_{2T} I_t^{1-\beta(\cdot,\cdot)} \partial_3 g[x]_\gamma^{\alpha,\beta}(t) \right) \right]_{t=T}$$

$$+ \left[\gamma_2 h_2(t) \left({}_T I_t^{1-\beta(\cdot,\cdot)} \partial_3 g[x]_\gamma^{\alpha,\beta}(t) - {}_a I_t^{1-\beta(\cdot,\cdot)} \partial_3 g[x]_\gamma^{\alpha,\beta}(t) \right) \right]_{t=b}.$$

Since (4.37) or (4.38) fails, there exists a function h_2 such that

$$\left. \frac{\partial i}{\partial \epsilon_2} \right|_{(0,0)} \neq 0.$$

In fact, if not, from the arbitrariness of the function h_2 and the fundamental lemma of the calculus of the variations, (4.37) and (4.38) would be verified. Thus, we may apply the implicit function theorem that ensures the existence of a function $\epsilon_2(\cdot)$

defined in a neighborhood of zero, such that $i(\epsilon_1, \epsilon_2(\epsilon_1)) = 0$. In conclusion, there exists a subfamily of variations of the form (4.42) that verifies the integral constraint (4.36). We now seek to prove the main result. For that purpose, consider the auxiliary function $j(\epsilon_1, \epsilon_2) = \mathcal{J}(x^*, T^*)$.

By hypothesis, function j attains a local minimum at $(0, 0)$ when subject to the constraint $i(\cdot, \cdot) = 0$, and we proved before that $\nabla i(0, 0) \neq (0, 0)$. Applying the Lagrange multiplier rule, we ensure the existence of a number λ such that

$$\nabla \left(j(0, 0) - \lambda i(0, 0) \right) = (0, 0).$$

In particular,

$$\frac{\partial (j - \lambda i)}{\partial \epsilon_1}(0, 0) = 0. \tag{4.45}$$

Let $F = L - \lambda g$. The relation (4.45) can be written as

$$
\begin{aligned}
0 = &\int_a^T h_1(t) \left[\partial_2 F[x]_\gamma^{\alpha,\beta}(t) + D_{\overline{\gamma}, T}^{\beta(\cdot, \cdot), \alpha(\cdot, \cdot)} \partial_3 F[x]_\gamma^{\alpha,\beta}(t) \right] dt \\
&+ \int_T^b \gamma_2 h_1(t) \left[{}_a D_t^{\beta(\cdot, \cdot)} \partial_3 F[x]_\gamma^{\alpha,\beta}(t) - {}_T D_t^{\beta(\cdot, \cdot)} \partial_3 F[x]_\gamma^{\alpha,\beta}(t) \right] dt \\
&+ h_1(T) \left[\gamma_1 \, {}_t I_T^{1-\alpha(\cdot, \cdot)} \partial_3 F[x]_\gamma^{\alpha,\beta}(t) - \gamma_2 \, {}_T I_t^{1-\beta(\cdot, \cdot)} \partial_3 F[x]_\gamma^{\alpha,\beta}(t) + \partial_2 \phi(t, x(t)) \right]_{t=T} \\
&+ \Delta T \left[F[x]_\gamma^{\alpha,\beta}(t) + \partial_1 \phi(t, x(t)) + \partial_2 \phi(t, x(t)) x'(t) + \lambda \psi'(t) \right]_{t=T} \\
&+ h_1(b) \gamma_2 \left[{}_T I_t^{1-\beta(\cdot, \cdot)} \partial_3 F[x]_\gamma^{\alpha,\beta}(t) - {}_a I_t^{1-\beta(\cdot, \cdot)} \partial_3 F[x]_\gamma^{\alpha,\beta}(t) \right]_{t=b}.
\end{aligned}
\tag{4.46}
$$

As h_1 and ΔT are arbitrary, we can choose $\Delta T = 0$ and $h_1(t) = 0$ for all $t \in [T, b]$. But h_1 is arbitrary in $t \in [a, T)$. Then, we obtain the first necessary condition (4.39):

$$\partial_2 F[x]_\gamma^{\alpha,\beta}(t) + D_{\overline{\gamma}, T}^{\beta(\cdot, \cdot), \alpha(\cdot, \cdot)} \partial_3 F[x]_\gamma^{\alpha,\beta}(t) = 0 \quad \forall t \in [a, T].$$

Analogously, considering $\Delta T = 0$ and $h_1(t) = 0$ for all $t \in [a, T] \cup \{b\}$, and h_1 arbitrary on (T, b), we obtain the second necessary condition (4.40):

$$\gamma_2 \left({}_a D_t^{\beta(\cdot, \cdot)} \partial_3 F[x]_\gamma^{\alpha,\beta}(t) - {}_T D_t^{\beta(\cdot, \cdot)} \partial_3 F[x]_\gamma^{\alpha,\beta}(t) \right) = 0 \quad \forall t \in [T, b].$$

As (x, T) is a solution to the necessary conditions (4.39) and (4.40), then Eq. (4.46) takes the form

$$
\begin{aligned}
&h_1(T) \left[\gamma_1 \, {}_t I_T^{1-\alpha(\cdot, \cdot)} \partial_3 F[x]_\gamma^{\alpha,\beta}(t) - \gamma_2 \, {}_T I_t^{1-\beta(\cdot, \cdot)} \partial_3 F[x]_\gamma^{\alpha,\beta}(t) + \partial_2 \phi(t, x(t)) \right]_{t=T} \\
&+ \Delta T \left[F[x]_\gamma^{\alpha,\beta}(t) + \partial_1 \phi(t, x(t)) + \partial_2 \phi(t, x(t)) x'(t) + \lambda \psi'(t) \right]_{t=T} \\
&+ h_1(b) \left[\gamma_2 \left({}_T I_t^{1-\beta(\cdot, \cdot)} \partial_3 F[x]_\gamma^{\alpha,\beta}(t) - {}_a I_t^{1-\beta(\cdot, \cdot)} \partial_3 F[x]_\gamma^{\alpha,\beta}(t) \right) \right]_{t=b} = 0.
\end{aligned}
\tag{4.47}
$$

Transversality conditions (4.41) are obtained for appropriate choices of variations.

In the next theorem, considering the same Problem 4, we rewrite the transversality conditions (4.41) in terms of the increment on time ΔT and on the increment of space Δx_T given by

$$\Delta x_T = (x + h_1)(T + \Delta T) - x(T). \tag{4.48}$$

Theorem 52 *Let* (x, T) *be a local minimizer to the functional* (4.33) *on* D *subject to the boundary condition* (4.35) *and the isoperimetric constraint* (4.36). *Then* (x, T) *satisfies the transversality conditions*

$$\begin{cases} F[x]_\gamma^{\alpha,\beta}(T) + \partial_1\phi(T, x(T)) + \lambda\psi'(T) \\ \quad +x'(T)\left[\gamma_{2T}I_t^{1-\beta(\cdot,\cdot)}\partial_3 F[x]_\gamma^{\alpha,\beta}(t) - \gamma_{1t}I_T^{1-\alpha(\cdot,\cdot)}\partial_3 F[x]_\gamma^{\alpha,\beta}(t)\right]_{t=T} = 0, \\ \left[\gamma_{1\,t}I_T^{1-\alpha(\cdot,\cdot)}\partial_3 F[x]_\gamma^{\alpha,\beta}(t) - \gamma_{2\,T}I_t^{1-\beta(\cdot,\cdot)}\partial_3 F[x]_\gamma^{\alpha,\beta}(t)\right]_{t=T} + \partial_2\phi(T, x(T)) = 0, \\ \gamma_2\left[{}_TI_t^{1-\beta(\cdot,\cdot)}\partial_3 F[x]_\gamma^{\alpha,\beta}(t) - {}_aI_t^{1-\beta(\cdot,\cdot)}\partial_3 F[x]_\gamma^{\alpha,\beta}(t)\right]_{t=b} = 0. \end{cases} \tag{4.49}$$

Proof Suppose (x^*, T^*) is an admissible variation of the form (4.42) with $\epsilon_1 = 1$ and $\epsilon_2 = 0$. Using Taylor's expansion up to first order for a small ΔT, and restricting the set of variations to those for which $h_1'(T) = 0$, we obtain the increment Δx_T on x:

$$(x + h_1)(T + \Delta T) = (x + h_1)(T) + x'(T)\Delta T + O(\Delta T)^2.$$

Relation (4.48) allows us to express $h_1(T)$ in terms of ΔT and Δx_T:

$$h_1(T) = \Delta x_T - x'(T)\Delta T + O(\Delta T)^2.$$

Substituting this expression into (4.47), and using appropriate choices of variations, we obtain the new transversality conditions (4.49).

Theorem 53 *Suppose that* (x, T) *gives a local minimum for functional* (4.33) *on* D *subject to the boundary condition* (4.35) *and the isoperimetric constraint* (4.36). *Then, there exists* $(\lambda_0, \lambda) \neq (0, 0)$ *such that if we define the function* $F : [a, b] \times \mathbb{R}^2 \to \mathbb{R}$ *by* $F = \lambda_0 L - \lambda g$, (x, T) *satisfies the following fractional Euler–Lagrange equations:*

$$\partial_2 F[x]_\gamma^{\alpha,\beta}(t) + D_{\bar\gamma,T}^{\beta(\cdot,\cdot),\alpha(\cdot,\cdot)}\partial_3 F[x]_\gamma^{\alpha,\beta}(t) = 0$$

on the interval $[a, T]$ *and*

$$\gamma_2\left({}_aD_t^{\beta(\cdot,\cdot)}\partial_3 F[x]_\gamma^{\alpha,\beta}(t) - {}_TD_t^{\beta(\cdot,\cdot)}\partial_3 F[x]_\gamma^{\alpha,\beta}(t)\right) = 0$$

on the interval $[T, b]$.

Proof If (x, T) does not verify (4.37) or (4.38), then the hypothesis of Theorem 51 is satisfied and we prove Theorem 53 considering $\lambda_0 = 1$. If (x, T) verifies (4.37) and (4.38), then we prove the result by considering $\lambda = 1$ and $\lambda_0 = 0$.

4.5.2 Necessary Optimality Conditions II

We now consider a new isoperimetric-type problem with the isoperimetric constraint of form

$$\int_a^b g[x]_\gamma^{\alpha,\beta}(t)dt = C, \tag{4.50}$$

where C is a given real number.

Problem 5 Determine the local minimizers of $\mathcal{J}$ (4.33), over all $(x, T) \in D$ satisfying a boundary condition

$$x(a) = x_a \tag{4.51}$$

for a fixed $x_a \in \mathbb{R}$ and an integral constraint of the form (4.50).

In the following theorem, we give fractional necessary optimality conditions for Problem 5.

Theorem 54 *Suppose that (x, T) gives a local minimum for functional (4.33) on D subject to the boundary condition (4.51) and the isoperimetric constraint (4.50). If (x, T) does not satisfy the Euler–Lagrange equation with respect to the isoperimetric constraint, that is, the condition*

$$\partial_2 g[x]_\gamma^{\alpha,\beta}(t) + D_{\overline{\gamma},b}^{\beta(\cdot,\cdot),\alpha(\cdot,\cdot)} \partial_3 g[x]_\gamma^{\alpha,\beta}(t) = 0, \quad t \in [a, b],$$

is not satisfied, then there exists $\lambda \neq 0$ such that if we define the function $F : [a, b] \times \mathbb{R}^2 \to \mathbb{R}$ by $F = L - \lambda g$, (x, T) satisfies the fractional Euler–Lagrange equations

$$\partial_2 F[x]_\gamma^{\alpha,\beta}(t) + D_{\overline{\gamma},T}^{\beta(\cdot,\cdot),\alpha(\cdot,\cdot)} \partial_3 L[x]_\gamma^{\alpha,\beta}(t) - \lambda D_{\overline{\gamma},b}^{\beta(\cdot,\cdot),\alpha(\cdot,\cdot)} \partial_3 g[x]_\gamma^{\alpha,\beta}(t) = 0 \tag{4.52}$$

on the interval $[a, T]$ and

$$\gamma_2 \left({}_a D_t^{\beta(\cdot,\cdot)} \partial_3 F[x]_\gamma^{\alpha,\beta}(t) - {}_T D_t^{\beta(\cdot,\cdot)} \partial_3 L[x]_\gamma^{\alpha,\beta}(t) \right)$$

$$-\lambda \left(\partial_2 g[x]_\gamma^{\alpha,\beta}(t) + \gamma_{1t} D_b^{\alpha(\cdot,\cdot)} \partial_3 g[x]_\gamma^{\alpha,\beta}(t) \right) = 0 \tag{4.53}$$

on the interval $[T, b]$. Moreover, (x, T) satisfies the transversality conditions

$$\begin{cases} L[x]_\gamma^{\alpha,\beta}(T) + \partial_1 \phi(T, x(T)) + \partial_2 \phi(T, x(T))x'(T) = 0, \\ \left[\gamma_1 {}_t I_T^{1-\alpha(\cdot,\cdot)} \partial_3 L[x]_\gamma^{\alpha,\beta}(t) - \gamma_2 {}_T I_t^{1-\beta(\cdot,\cdot)} \partial_3 L[x]_\gamma^{\alpha,\beta}(t) + \partial_2 \phi(t, x(t)) \right]_{t=T} = 0 \\ \left[-\lambda \gamma_{1t} I_b^{1-\alpha(\cdot,\cdot)} \partial_3 g[x]_\gamma^{\alpha,\beta}(t) \right. \\ \left. + \gamma_2 \left({}_T I_t^{1-\beta(\cdot,\cdot)} \partial_3 L[x]_\gamma^{\alpha,\beta}(t) - {}_a I_t^{1-\beta(\cdot,\cdot)} \partial_3 F[x]_\gamma^{\alpha,\beta}(t) \right) \right]_{t=b} = 0. \end{cases}$$

$$\tag{4.54}$$

Proof Similarly as done to prove Theorem 51, let

$$(x^*, T^*) = (x + \epsilon_1 h_1 + \epsilon_2 h_2, T + \epsilon_1 \Delta T)$$

be a variation of the solution, and define

$$i(\epsilon_1, \epsilon_2) = \int_a^b g[x^*]_\gamma^{\alpha,\beta}(t) dt - C.$$

The derivative $\dfrac{\partial i}{\partial \epsilon_2}$, when $\epsilon_1 = \epsilon_2 = 0$, is

$$\left. \frac{\partial i}{\partial \epsilon_2} \right|_{(0,0)} = \int_a^b \left(\partial_2 g[x]_\gamma^{\alpha,\beta}(t) h_2(t) + \partial_3 g[x]_\gamma^{\alpha,\beta}(t)^C D_\gamma^{\alpha(\cdot,\cdot),\beta(\cdot,\cdot)} h_2(t) \right) dt.$$

Integrating by parts and choosing variations such that $h_2(b) = 0$, we have

$$\left. \frac{\partial i}{\partial \epsilon_2} \right| (0,0) = \int_a^b h_2(t) \left[\partial_2 g[x]_\gamma^{\alpha,\beta}(t) + D_{\overline{\gamma},b}^{\beta(\cdot,\cdot),\alpha(\cdot,\cdot)} \partial_3 g[x]_\gamma^{\alpha,\beta}(t) \right] dt.$$

Thus, there exists a function h_2 such that

$$\left. \frac{\partial i}{\partial \epsilon_2} \right| (0,0) \neq 0.$$

We may apply the implicit function theorem to conclude that there exists a subfamily of variations satisfying the integral constraint. Consider the new function $j(\epsilon_1, \epsilon_2) = \mathcal{J}(x^*, T^*)$. Since j has a local minimum at $(0,0)$ when subject to the constraint $i(\cdot, \cdot) = 0$ and $\nabla i(0,0) \neq (0,0)$, there exists a number λ such that

$$\frac{\partial}{\partial \epsilon_1} (j - \lambda i)(0,0) = 0. \tag{4.55}$$

Let $F = L - \lambda g$. Relation (4.55) can be written as

$$\int_a^T h_1(t) \left[\partial_2 F[x]_\gamma^{\alpha,\beta}(t) + D_{\overline{\gamma},T}^{\beta(\cdot,\cdot),\alpha(\cdot,\cdot)} \partial_3 L[x]_\gamma^{\alpha,\beta}(t) - \lambda D_{\overline{\gamma},b}^{\beta(\cdot,\cdot),\alpha(\cdot,\cdot)} \partial_3 g[x]_\gamma^{\alpha,\beta}(t) \right] dt$$

$$+ \int_T^b h_1(t) \left[\gamma_2 \left({}_a D_t^{\beta(\cdot,\cdot)} \partial_3 F[x]_\gamma^{\alpha,\beta}(t) - {}_T D_t^{\beta(\cdot,\cdot)} \partial_3 L[x]_\gamma^{\alpha,\beta}(t) \right) \right.$$

$$\left. - \lambda \left(\partial_2 g[x]_\gamma^{\alpha,\beta}(t) + \gamma_{1t} D_b^{\alpha(\cdot,\cdot)} \partial_3 g[x]_\gamma^{\alpha,\beta}(t) \right) \right] dt$$

$$+ h_1(T) \left[\gamma_1 \, {}_t I_T^{1-\alpha(\cdot,\cdot)} \partial_3 L[x]_\gamma^{\alpha,\beta}(t) - \gamma_2 \, {}_T I_t^{1-\beta(\cdot,\cdot)} \partial_3 L[x]_\gamma^{\alpha,\beta}(t) + \partial_2 \phi(t, x(t)) \right]_{t=T}$$

$$+ \Delta T \left[L[x]_\gamma^{\alpha,\beta}(t) + \partial_1 \phi(t, x(t)) + \partial_2 \phi(t, x(t)) x'(t) \right]_{t=T}$$

$$+ h_1(b) \left[-\lambda \gamma_{1t} I_b^{1-\alpha(\cdot,\cdot)} \partial_3 g[x]_\gamma^{\alpha,\beta}(t) \right.$$
$$\left. + \gamma_2 \left({}_T I_t^{1-\beta(\cdot,\cdot)} \partial_3 L[x]_\gamma^{\alpha,\beta}(t) - {}_a I_t^{1-\beta(\cdot,\cdot)} \partial_3 F[x]_\gamma^{\alpha,\beta}(t) \right) \right]_{t=b} = 0.$$

Considering appropriate choices of variations, we obtain the first (4.52) and the second (4.53) necessary optimality conditions and also the transversality conditions (4.54).

Similarly to Theorem 53, the following result holds.

Theorem 55 *Suppose that* (x, T) *gives a local minimum for functional* (4.33) *on* D *subject to the boundary condition* (4.51) *and the isoperimetric constraint* (4.50). *Then, there exists* $(\lambda_0, \lambda) \neq (0, 0)$ *such that if we define the function* $F : [a, b] \times \mathbb{R}^2 \to \mathbb{R}$ *by* $F = \lambda_0 L - \lambda g$, (x, T) *satisfies the fractional Euler–Lagrange equations*

$$\partial_2 F[x]_\gamma^{\alpha,\beta}(t) + D_{\overline{\gamma},T}^{\beta(\cdot,\cdot),\alpha(\cdot,\cdot)} \partial_3 L[x]_\gamma^{\alpha,\beta}(t) - \lambda D_{\overline{\gamma},b}^{\beta(\cdot,\cdot),\alpha(\cdot,\cdot)} \partial_3 g[x]_\gamma^{\alpha,\beta}(t) = 0$$

on the interval $[a, T]$ *and*

$$\gamma_2 \left({}_a D_t^{\beta(\cdot,\cdot)} \partial_3 F[x]_\gamma^{\alpha,\beta}(t) - {}_T D_t^{\beta(\cdot,\cdot)} \partial_3 L[x]_\gamma^{\alpha,\beta}(t) \right)$$
$$- \lambda \left(\partial_2 g[x]_\gamma^{\alpha,\beta}(t) + \gamma_{1t} D_b^{\alpha(\cdot,\cdot)} \partial_3 g[x]_\gamma^{\alpha,\beta}(t) \right) = 0$$

on the interval $[T, b]$.

4.5.3 Example

Let $\alpha(t, \tau) = \alpha(t)$ and $\beta(t, \tau) = \beta(\tau)$. Define the function

$$\psi(T) = \int_0^T \left(\frac{t^{1-\alpha(t)}}{2\Gamma(2 - \alpha(t))} + \frac{(b-t)^{1-\beta(t)}}{2\Gamma(2 - \beta(t))} \right)^2 dt$$

on the interval $[0, b]$ with $b > 0$. Consider the functional J defined by

$$J(x, t) = \int_0^T \left[\alpha(t) + \left({}^C D_\gamma^{\alpha(\cdot,\cdot),\beta(\cdot,\cdot)} x(t) \right)^2 \right.$$
$$\left. + \left(\frac{t^{1-\alpha(t)}}{2\Gamma(2 - \alpha(t))} + \frac{(b-t)^{1-\beta(t)}}{2\Gamma(2 - \beta(t))} \right)^2 \right] dt$$

for $t \in [0, b]$ and $\gamma = (1/2, 1/2)$, subject to the initial condition

$$x(0) = 0$$

and the isoperimetric constraint

$$\int_0^T {}^C D_\gamma^{\alpha(\cdot,\cdot),\beta(\cdot,\cdot)} x(t) \left(\frac{t^{1-\alpha(t)}}{2\Gamma(2-\alpha(t))} + \frac{(b-t)^{1-\beta(t)}}{2\Gamma(2-\beta(t))} \right)^2 dt = \psi(T).$$

Define $F = L - \lambda g$ with $\lambda = 2$, that is,

$$F = \alpha(t) + \left({}^C D_\gamma^{\alpha(\cdot),\beta(\cdot)} x(t) - \frac{t^{1-\alpha(t)}}{2\Gamma(2-\alpha(t))} - \frac{(b-t)^{1-\beta(t)}}{2\Gamma(2-\beta(t))} \right)^2.$$

Consider the function $\overline{x}(t) = t$ with $t \in [0, b]$. Because

$$ {}^C D_\gamma^{\alpha(\cdot,\cdot),\beta(\cdot,\cdot)} \overline{x}(t) = \frac{t^{1-\alpha(t)}}{2\Gamma(2-\alpha(t))} + \frac{(b-t)^{1-\beta(t)}}{2\Gamma(2-\beta(t))},$$

we have that $\overline{x}$ satisfies conditions (4.39), (4.40) and the two last of (4.41). Using the first condition of (4.41), that is,

$$\alpha(t) + 2 \left(\frac{T^{1-\alpha(T)}}{2\Gamma(2-\alpha(T))} + \frac{(b-T)^{1-\beta(T)}}{2\Gamma(2-\beta(T))} \right)^2 = 0,$$

we obtain the optimal time T.

4.6   Variational Problems with Holonomic Constraints

In this section, we present a new variational problem subject to a new type of constraints. A holonomic constraint is a condition of the form

$$g(t, x) = 0,$$

where $x = (x_1, x_2, ..., x_n), n \geq 2$ and g is a given function (see, e.g., van Brunt [25]).
   Consider the space

$$U = \{(x_1, x_2, T) \in C^1([a, b]) \times C^1([a, b]) \times [a, b] : x_1(a) = x_{1a} \wedge x_2(a) = x_{2a}\} \tag{4.56}$$

for fixed reals $x_{1a}, x_{2a} \in \mathbb{R}$. In this section, we consider the following variational problem:

Problem 6 Find functions x_1 and x_2 that maximize or minimize the functional $\mathcal{J}$ defined in U by

$$\mathcal{J}(x_1, x_2, T) = \int_a^T L(t, x_1(t), x_2(t), {}^C D_\gamma^{\alpha(\cdot,\cdot),\beta(\cdot,\cdot)} x_1(t), {}^C D_\gamma^{\alpha(\cdot,\cdot),\beta(\cdot,\cdot)} x_2(t)) dt$$
$$+ \phi(T, x_1(T), x_2(T)),$$

$$(4.57)$$

where the admissible functions satisfy the constraint

$$g(t, x_1(t), x_2(t)) = 0, \quad t \in [a, b], \qquad (4.58)$$

called a holonomic constraint, where $g : [a, b] \times \mathbb{R}^2 \to \mathbb{R}$ is a continuous function and continuously differentiable with respect to second and third arguments.

The terminal time T and terminal states $x_1(T)$ and $x_2(T)$ are free, and the Lagrangian $L : [a, b] \times \mathbb{R}^4 \to \mathbb{R}$ is a continuous function and continuously differentiable with respect to its ith argument, $i \in \{2, 3, 4, 5\}$. The terminal cost function $\phi : [a, b] \times \mathbb{R}^2 \to \mathbb{R}$ is of class C^1.

4.6.1 Necessary Optimality Conditions

The next theorem gives fractional necessary optimality conditions to the variational problem with a holonomic constraint. To simplify the notation, we denote by x the vector (x_1, x_2); by ${}^C D_\gamma^{\alpha(\cdot,\cdot),\beta(\cdot,\cdot)} x$, we mean the two-dimensional vector $({}^C D_\gamma^{\alpha(\cdot,\cdot),\beta(\cdot,\cdot)} x_1, {}^C D_\gamma^{\alpha(\cdot,\cdot),\beta(\cdot,\cdot)} x_2)$; we use the operator

$$[x]_\gamma^{\alpha,\beta}(t) := \left(t, x(t), {}^C D_\gamma^{\alpha(\cdot,\cdot),\beta(\cdot,\cdot)} x(t)\right).$$

The next result gives necessary optimality conditions for Problem 6.

Theorem 56 *Suppose that (x, T) gives a local minimum to functional $\mathcal{J}$ as in (4.57), under the constraint (4.58) and the boundary conditions defined in (4.56). If*

$$\partial_3 g(t, x(t)) \neq 0 \ \forall t \in [a, b],$$

then there exists a piecewise continuous function $\lambda : [a, b] \to \mathbb{R}$ such that (x, T) satisfies the following fractional Euler–Lagrange equations:

$$\partial_2 L[x]_\gamma^{\alpha,\beta}(t) + D_{\overline{\gamma},T}^{\beta(\cdot,\cdot),\alpha(\cdot,\cdot)} \partial_4 L[x]_\gamma^{\alpha,\beta}(t) + \lambda(t)\partial_2 g(t, x(t)) = 0 \qquad (4.59)$$

and

$$\partial_3 L[x]_\gamma^{\alpha,\beta}(t) + D_{\overline{\gamma},T}^{\beta(\cdot,\cdot),\alpha(\cdot,\cdot)} \partial_5 L[x]_\gamma^{\alpha,\beta}(t) + \lambda(t)\partial_3 g(t, x(t)) = 0 \qquad (4.60)$$

on the interval $[a, T]$ and

$$\gamma_2 \left({}_a D_t^{\beta(\cdot,\cdot)} \partial_4 L[x]_\gamma^{\alpha,\beta}(t) - {}_T D_t^{\beta(\cdot,\cdot)} \partial_4 L[x]_\gamma^{\alpha,\beta}(t) + \lambda(t)\partial_2 g(t, x(t)) \right) = 0 \quad (4.61)$$

and

$$_aD_t^{\beta(\cdot,\cdot)}\partial_5 L[x]_\gamma^{\alpha,\beta}(t) - {_TD_t^{\beta(\cdot,\cdot)}}\partial_5 L[x]_\gamma^{\alpha,\beta}(t) + \lambda(t)\partial_3 g(t, x(t)) = 0 \qquad (4.62)$$

on the interval $[T, b]$. *Moreover,* (x, T) *satisfies the transversality conditions*

$$
\begin{cases}
L[x]_\gamma^{\alpha,\beta}(T) + \partial_1\phi(T, x(T)) + \partial_2\phi(T, x(T))x_1'(T) + \partial_3\phi(T, x(T))x_2'(T) = 0, \\
\left[\gamma_1{_tI_T^{1-\alpha(\cdot,\cdot)}}\partial_4 L[x]_\gamma^{\alpha,\beta}(t) - \gamma_2{_TI_t^{1-\beta(\cdot,\cdot)}}\partial_4 L[x]_\gamma^{\alpha,\beta}(t)\right]_{t=T} + \partial_2\phi(T, x(T)) = 0, \\
\left[\gamma_{1t}I_T^{1-\alpha(\cdot,\cdot)}\partial_5 L[x]_\gamma^{\alpha,\beta}(t) - \gamma_{2T}I_t^{1-\beta(\cdot,\cdot)}\partial_5 L[x]_\gamma^{\alpha,\beta}(t)\right]_{t=T} + \partial_3\phi(T, x(T)) = 0, \\
\gamma_2\left[{_TI_t^{1-\beta(\cdot,\cdot)}}\partial_4 L[x]_\gamma^{\alpha,\beta}(t) - {_aI_t^{1-\beta(\cdot,\cdot)}}\partial_4 L[x]_\gamma^{\alpha,\beta}(t)\right]_{t=b} = 0, \\
\gamma_2\left[{_TI_t^{1-\beta(\cdot,\cdot)}}\partial_5 L[x]_\gamma^{\alpha,\beta}(t) - {_aI_t^{1-\beta(\cdot,\cdot)}}\partial_5 L[x]_\gamma^{\alpha,\beta}(t)\right]_{t=b} = 0.
\end{cases}
$$

$$(4.63)$$

Proof Consider admissible variations of the optimal solution (x, T) of the type

$$(x^*, T^*) = (x + \epsilon h, T + \epsilon \Delta T),$$

where $\epsilon \in \mathbb{R}$ is a small parameter, $h = (h_1, h_2) \in C^1([a, b]) \times C^1([a, b])$ satisfies $h_i(a) = 0, i = 1, 2$, and $\Delta T \in \mathbb{R}$. Because

$$\partial_3 g(t, x(t)) \neq 0 \quad \forall t \in [a, b],$$

by the implicit function theorem there exists a subfamily of variations of (x, T) that satisfy (4.58); that is, there exists a unique function $h_2(\epsilon, h_1)$ such that the admissible variation (x^*, T^*) satisfies the holonomic constraint (4.58):

$$g(t, x_1(t) + \epsilon h_1(t), x_2(t) + \epsilon h_2(t)) = 0 \quad \forall t \in [a, b].$$

Differentiating this condition with respect to ϵ and considering $\epsilon = 0$, we obtain that

$$\partial_2 g(t, x(t))h_1(t) + \partial_3 g(t, x(t))h_2(t) = 0,$$

which is equivalent to

$$\frac{\partial_2 g(t, x(t))h_1(t)}{\partial_3 g(t, x(t))} = -h_2(t). \qquad (4.64)$$

Define j on a neighborhood of zero by

$$j(\epsilon) = \int_a^{T+\epsilon\Delta T} L[x^*]_\gamma^{\alpha,\beta}(t)dt + \phi(T + \epsilon\Delta T, x^*(T + \epsilon\Delta T)).$$

The derivative $\dfrac{\partial j}{\partial \epsilon}$ for $\epsilon = 0$ is

$$\frac{\partial j}{\partial \epsilon}\bigg|_{\epsilon=0} = \int_a^T \left(\partial_2 L[x]_\gamma^{\alpha,\beta}(t) h_1(t) + \partial_3 L[x]_\gamma^{\alpha,\beta}(t) h_2(t) \right.$$

$$+ \partial_4 L[x]_\gamma^{\alpha,\beta}(t)\, {}^C D_\gamma^{\alpha(\cdot,\cdot),\beta(\cdot,\cdot)} h_1(t) + \partial_5 L[x]_\gamma^{\alpha,\beta}(t)\, {}^C D_\gamma^{\alpha(\cdot,\cdot),\beta(\cdot,\cdot)} h_2(t) \bigg) dt$$

$$+ L[x]_\gamma^{\alpha,\beta}(T)\Delta T + \partial_1 \phi(T, x(T))\Delta T + \partial_2 \phi(T, x(T)) \left[h_1(T) + x_1'(T)\Delta T \right]$$

$$+ \partial_3 \phi(T, x(T)) \left[h_2(T) + x_2'(T)\Delta T \right].$$

$$(4.65)$$

The third term in (4.65) can be written as

$$\int_a^T \partial_4 L[x]_\gamma^{\alpha,\beta}(t)\, {}^C D_\gamma^{\alpha(\cdot,\cdot),\beta(\cdot,\cdot)} h_1(t)\, dt$$

$$= \int_a^T \partial_4 L[x]_\gamma^{\alpha,\beta}(t) \left[\gamma_1\, {}^C_a D_t^{\alpha(\cdot,\cdot)} h_1(t) + \gamma_2\, {}^C_t D_b^{\beta(\cdot,\cdot)} h_1(t) \right] dt$$

$$= \gamma_1 \int_a^T \partial_4 L[x]_\gamma^{\alpha,\beta}(t)\, {}^C_a D_t^{\alpha(\cdot,\cdot)} h_1(t)\, dt$$

$$+ \gamma_2 \left[\int_a^b \partial_4 L[x]_\gamma^{\alpha,\beta}(t)\, {}^C_t D_b^{\beta(\cdot,\cdot)} h_1(t)\, dt - \int_T^b \partial_4 L[x]_\gamma^{\alpha,\beta}(t)\, {}^C_t D_b^{\beta(\cdot,\cdot)} h_1(t)\, dt \right].$$

$$(4.66)$$

Integrating by parts, (4.66) can be written as

$$\int_a^T h_1(t) \left[\gamma_{1t} D_T^{\alpha(\cdot,\cdot)} \partial_4 L[x]_\gamma^{\alpha,\beta}(t) + \gamma_{2a} D_t^{\beta(\cdot,\cdot)} \partial_4 L[x]_\gamma^{\alpha,\beta}(t) \right] dt$$

$$+ \int_T^b \gamma_2 h_1(t) \left[{}_a D_t^{\beta(\cdot,\cdot)} \partial_4 L[x]_\gamma^{\alpha,\beta}(t) - {}_T D_t^{\beta(\cdot,\cdot)} \partial_4 L[x]_\gamma^{\alpha,\beta}(t) \right] dt$$

$$+ \left[h_1(t) \left(\gamma_{1t} I_T^{1-\alpha(\cdot,\cdot)} \partial_4 L[x]_\gamma^{\alpha,\beta}(t) - \gamma_{2T} I_t^{1-\beta(\cdot,\cdot)} \partial_4 L[x]_\gamma^{\alpha,\beta}(t) \right) \right]_{t=T}$$

$$+ \left[\gamma_2 h_1(t) \left({}_T I_t^{1-\beta(\cdot,\cdot)} \partial_4 L[x]_\gamma^{\alpha,\beta}(t) - {}_a I_t^{1-\beta(\cdot,\cdot)} \partial_4 L[x]_\gamma^{\alpha,\beta}(t) \right) \right]_{t=b}.$$

By proceeding similarly to the fourth term in (4.65), we obtain an equivalent expression. Substituting these relations into (4.65) and considering the fractional operator $D_{\overline{\gamma},c}^{\beta(\cdot,\cdot),\alpha(\cdot,\cdot)}$ as defined in (4.34), we obtain that

$$0 = \int_a^T \left[h_1(t) \left[\partial_2 L[x]_\gamma^{\alpha,\beta}(t) + D_{\overline{\gamma},T}^{\beta(\cdot,\cdot),\alpha(\cdot,\cdot)} \partial_4 L[x]_\gamma^{\alpha,\beta}(t) \right] \right.$$

$$+ h_2(t) \left[\partial_3 L[x]_\gamma^{\alpha,\beta}(t) + D_{\overline{\gamma},T}^{\beta(\cdot,\cdot),\alpha(\cdot,\cdot)} \partial_5 L[x]_\gamma^{\alpha,\beta}(t) \right] \bigg] dt$$

$$+ \gamma_2 \int_T^b \left[h_1(t) \left[{}_a D_t^{\beta(\cdot,\cdot)} \partial_4 L[x]_\gamma^{\alpha,\beta}(t) - {}_T D_t^{\beta(\cdot,\cdot)} \partial_4 L[x]_\gamma^{\alpha,\beta}(t) \right] \right.$$

$$+ h_2(t) \left[{}_a D_t^{\beta(\cdot,\cdot)} \partial_5 L[x]_\gamma^{\alpha,\beta}(t) - {}_T D_t^{\beta(\cdot,\cdot)} \partial_5 L[x]_\gamma^{\alpha,\beta}(t) \right] dt \right]$$

$$+ h_1(T) \left[\gamma_{1t} I_T^{1-\alpha(\cdot,\cdot)} \partial_4 L[x]_\gamma^{\alpha,\beta}(t) \right.$$

$$\left. - \gamma_{2T} I_t^{1-\beta(\cdot,\cdot)} \partial_4 L[x]_\gamma^{\alpha,\beta}(t) + \partial_2\phi(t,x(t)) \right]_{t=T} \qquad (4.67)$$

$$+ h_2(T) \left[\gamma_{1t} I_T^{1-\alpha(\cdot,\cdot)} \partial_5 L[x]_\gamma^{\alpha,\beta}(t) - \gamma_{2T} I_t^{1-\beta(\cdot,\cdot)} \partial_5 L[x]_\gamma^{\alpha,\beta}(t) + \partial_3\phi(t,x(t)) \right]_{t=T}$$

$$+ \Delta T \left[L[x]_\gamma^{\alpha,\beta}(t) + \partial_1\phi(t,x(t)) + \partial_2\phi(t,x(t))x_1'(t) + \partial_3\phi(t,x(t))x_2'(t) \right]_{t=T}$$

$$+ h_1(b) \left[\gamma_2 \left({}_T I_t^{1-\beta(\cdot,\cdot)} \partial_4 L[x]_\gamma^{\alpha,\beta}(t) - {}_a I_t^{1-\beta(\cdot,\cdot)} \partial_4 L[x]_\gamma^{\alpha,\beta}(t) \right) \right]_{t=b}$$

$$+ h_2(b) \left[\gamma_2 \left({}_T I_t^{1-\beta(\cdot,\cdot)} \partial_5 L[x]_\gamma^{\alpha,\beta}(t) - {}_a I_t^{1-\beta(\cdot,\cdot)} \partial_5 L[x]_\gamma^{\alpha,\beta}(t) \right) \right]_{t=b}.$$

Define the piecewise continuous function λ by

$$\lambda(t) = \begin{cases} -\dfrac{\partial_3 L[x]_\gamma^{\alpha,\beta}(t) + D_{\overline{\gamma},T}^{\beta(\cdot,\cdot),\alpha(\cdot,\cdot)} \partial_5 L[x]_\gamma^{\alpha,\beta}(t)}{\partial_3 g(t,x(t))}, & t \in [a,T] \\[3mm] -\dfrac{{}_a D_t^{\beta(\cdot,\cdot)} \partial_5 L[x]_\gamma^{\alpha,\beta}(t) - {}_T D_t^{\beta(\cdot,\cdot)} \partial_5 L[x]_\gamma^{\alpha,\beta}(t)}{\partial_3 g(t,x(t))}, & t \in [T,b]. \end{cases} \qquad (4.68)$$

Using Eqs. (4.64) and (4.68), we obtain that

$$\lambda(t)\partial_2 g(t,x(t))h_1(t)$$

$$= \begin{cases} (\partial_3 L[x]_\gamma^{\alpha,\beta}(t) + D_{\overline{\gamma},T}^{\beta(\cdot,\cdot),\alpha(\cdot,\cdot)} \partial_5 L[x]_\gamma^{\alpha,\beta}(t))h_2(t), & t \in [a,T] \\[2mm] ({}_a D_t^{\beta(\cdot,\cdot)} \partial_5 L[x]_\gamma^{\alpha,\beta}(t) - {}_T D_t^{\beta(\cdot,\cdot)} \partial_5 L[x]_\gamma^{\alpha,\beta}(t))h_2(t), & t \in [T,b]. \end{cases}$$

Substituting in (4.67), we have

$$0 = \int_a^T h_1(t) \left[\partial_2 L[x]_\gamma^{\alpha,\beta}(t) + D_{\overline{\gamma},T}^{\beta(\cdot,\cdot),\alpha(\cdot,\cdot)} \partial_4 L[x]_\gamma^{\alpha,\beta}(t) + \lambda(t)\partial_2 g(t,x(t)) \right] dt$$

$$+ \gamma_2 \int_T^b h_1(t) \left[{}_a D_t^{\beta(\cdot,\cdot)} \partial_4 L[x]_\gamma^{\alpha,\beta}(t) - {}_T D_t^{\beta(\cdot,\cdot)} \partial_4 L[x]_\gamma^{\alpha,\beta}(t) + \lambda(t)\partial_2 g(t,x(t)) \right] dt$$

$$+ h_1(T) \left[\gamma_{1t} I_T^{1-\alpha(\cdot,\cdot)} \partial_4 L[x]_\gamma^{\alpha,\beta}(t) - \gamma_{2T} I_t^{1-\beta(\cdot,\cdot)} \partial_4 L[x]_\gamma^{\alpha,\beta}(t) + \partial_2\phi(t,x(t)) \right]_{t=T}$$

$$+ h_2(T) \left[\gamma_{1t} I_T^{1-\alpha(\cdot,\cdot)} \partial_5 L[x]_\gamma^{\alpha,\beta}(t) - \gamma_{2T} I_t^{1-\beta(\cdot,\cdot)} \partial_5 L[x]_\gamma^{\alpha,\beta}(t) + \partial_3\phi(t,x(t)) \right]_{t=T}$$

$$+\Delta T\left[L[x]_\gamma^{\alpha,\beta}(t)+\partial_1\phi(t,x(t))+\partial_2\phi(t,x(t))x_1'(t)+\partial_3\phi(t,x(t))x_2'(t)\right]_{t=T}$$

$$+h_1(b)\left[\gamma_2\left({}_TI_t^{1-\beta(\cdot,\cdot)}\partial_4L[x]_\gamma^{\alpha,\beta}(t)-{}_aI_t^{1-\beta(\cdot,\cdot)}\partial_4L[x]_\gamma^{\alpha,\beta}(t)\right)\right]_{t=b}$$

$$+h_2(b)\left[\gamma_2\left({}_TI_t^{1-\beta(\cdot,\cdot)}\partial_5L[x]_\gamma^{\alpha,\beta}(t)-{}_aI_t^{1-\beta(\cdot,\cdot)}\partial_5L[x]_\gamma^{\alpha,\beta}(t)\right)\right]_{t=b}.$$

Considering appropriate choices of variations, we obtained the first (4.59) and the third (4.61) necessary conditions and also the transversality conditions (4.63). The remaining conditions (4.60) and (4.62) follow directly from (4.68).

4.6.2 Example

We end this section with a simple illustrative example. Consider the following problem:

$$J(x,t)=\int_0^T\left[\alpha(t)+\left({}^CD_\gamma^{\alpha(\cdot,\cdot),\beta(\cdot,\cdot)}x_1(t)-\frac{t^{1-\alpha(t)}}{2\Gamma(2-\alpha(t))}-\frac{(b-t)^{1-\beta(t)}}{2\Gamma(2-\beta(t))}\right)^2\right.$$
$$\left.+\left({}^CD_\gamma^{\alpha(\cdot,\cdot),\beta(\cdot,\cdot)}x_2(t)\right)^2\right]dt\longrightarrow\min,$$
$$x_1(t)+x_2(t)=t+1,$$
$$x_1(0)=0,\quad x_2(0)=1.$$

It is a simple exercise to check that $x_1(t)=t$, $x_2(t)\equiv1$ and $\lambda(t)\equiv0$ satisfy our Theorem 56.

4.7 Fractional Variational Herglotz Problem

In this section, we study fractional variational problems of Herglotz type, depending on the combined Caputo fractional derivatives ${}^CD_\gamma^{\alpha(\cdot,\cdot),\beta(\cdot,\cdot)}$. Two different cases are considered.

The variational problem of Herglotz is a generalization of the classical variational problem. It allows us to describe nonconservative processes, even in case the Lagrange function is autonomous (i.e., when the Lagrangian does not depend explicitly on time). In opposite to calculus of variations, where the cost functional is given by an integral depending only on time, space, and the dynamics, in the Herglotz variational problem the model is given by a differential equation involving the derivative of the objective function z, and the Lagrange function depends on time, trajectories

x and z, and the derivative of x. The problem of Herglotz was posed by Herglotz [11], but only in 1996, with the works [9, 10], it has gained the attention of the mathematical community. Since then, several papers were devoted to this subject. For example, see references Almeida and Malinowska [2], Georgieva and Guenther [7], Georgieva et al. [8], Santos et al. [17–19].

In Sect. 4.7.1, we obtain fractional Euler–Lagrange conditions for the fractional variational problem of Herglotz, with one variable, and the general case for several independent variables is discussed in Sect. 4.7.2. Finally, three illustrative examples are presented in detail (Sect. 4.7.3).

4.7.1 Fundamental Problem of Herglotz

Let $\alpha, \beta : [a, b]^2 \to (0, 1)$ be two functions. The fractional Herglotz variational problem that we study is as follows.

Problem 7 Determine the trajectories $x \in C^1 ([a, b]; \mathbb{R})$ satisfying a given boundary condition $x(a) = x_a$, for a fixed $x_a \in \mathbb{R}$, and a real $T \in (a, b]$, that extremize the value of $z(T)$, where z satisfies the differential equation

$$\dot{z}(x, t) = L\left(t, x(t), {}^C D_\gamma^{\alpha(\cdot,\cdot), \beta(\cdot,\cdot)} x(t), z(t)\right), \quad t \in [a, b], \qquad (4.69)$$

with dependence on a combined Caputo fractional derivative operator, subject to the initial condition

$$z(a) = z_a, \qquad (4.70)$$

where z_a is a fixed real number.

In the sequel, we use the auxiliary notation

$$[x, z]_\gamma^{\alpha, \beta}(t) = \left(t, x(t), {}^C D_\gamma^{\alpha(\cdot,\cdot), \beta(\cdot,\cdot)} x(t), z(t)\right).$$

The Lagrangian L is assumed to satisfy the following hypothesis:

1. $L : [a, b] \times \mathbb{R}^3 \to \mathbb{R}$ is differentiable.
2. $t \to \lambda(t) \partial_3 L[x, z]_\gamma^{\alpha, \beta}(t)$ is such that ${}_T D_t^{\beta(\cdot,\cdot)} \left(\lambda(t) \partial_3 L[x, z]_\gamma^{\alpha, \beta}(t)\right)$, ${}_a D_t^{\beta(\cdot,\cdot)} \left(\lambda(t) \partial_3 L[x, z]_\gamma^{\alpha, \beta}(t)\right)$, and $D_{\overline{\gamma}}^{\beta(\cdot,\cdot), \alpha(\cdot,\cdot)} \left(\lambda(t) \partial_3 L[x, z]_\gamma^{\alpha, \beta}(t)\right)$ exist and are continuous on $[a, b]$, where

$$\lambda(t) = \exp\left(-\int_a^t \partial_4 L\, [x, z]_\gamma^{\alpha, \beta}(\tau) d\tau\right).$$

The following result gives necessary conditions of Euler–Lagrange type for a solution of Problem 7.

Theorem 57 *Let $x \in C^1([a, b]; \mathbb{R})$ be such that z defined by Eq. (4.69), subject to the initial condition (4.70), has an extremum at $T \in]a, b]$. Then, (x, z) satisfies the fractional differential equations*

$$\partial_2 L[x, z]_\gamma^{\alpha,\beta}(t)\lambda(t) + D_{\overline{\gamma}}^{\beta(\cdot,\cdot),\alpha(\cdot,\cdot)} \left(\lambda(t)\partial_3 L[x, z]_\gamma^{\alpha,\beta}(t)\right) = 0, \qquad (4.71)$$

on $[a, T]$ and

$$\gamma_2 \left(_a D_t^{\beta(\cdot,\cdot)} \left(\lambda(t)\partial_3 L[x, z]_\gamma^{\alpha,\beta}(t)\right) - _T D_t^{\beta(\cdot,\cdot)} \left(\lambda(t)\partial_3 L[x, z]_\gamma^{\alpha,\beta}(t)\right)\right) = 0, \quad (4.72)$$

on $[T, b]$. Moreover, the following transversality conditions are satisfied:

$$\begin{cases} \left[\gamma_{1t} I_T^{1-\alpha(\cdot,\cdot)} \left(\lambda(t)\partial_3 L[x, z]_\gamma^{\alpha,\beta}(t)\right) - \gamma_{2T} I_t^{1-\beta(\cdot,\cdot)} \left(\lambda(t)\partial_3 L[x, z]_\gamma^{\alpha,\beta}(t)\right)\right]_{t=T} = 0, \\ \gamma_2 \left[_T I_t^{1-\beta(\cdot,\cdot)} \left(\lambda(t)\partial_3 L[x, z]_\gamma^{\alpha,\beta}(t)\right) - _a I_t^{1-\beta(\cdot,\cdot)} \left(\lambda(t)\partial_3 L[x, z]_\gamma^{\alpha,\beta}(t)\right)\right]_{t=b} = 0. \end{cases}$$
$$(4.73)$$

If $T < b$, then $L[x, z]_\gamma^{\alpha,\beta}(T) = 0$.

Proof Let x be a solution to the problem. Consider an admissible variation of x, $\overline{x} = x + \epsilon h$, where $h \in C^1([a, b]; \mathbb{R})$ is an arbitrary perturbation curve and $\epsilon \in \mathbb{R}$ represents a small number ($|\epsilon| \ll 1$). The constraint $x(a) = x_a$ implies that all admissible variations must fulfill the condition $h(a) = 0$. On the other hand, consider an admissible variation of z, $\overline{z} = z + \epsilon\theta$, where θ is a perturbation curve (not arbitrary) such that

1. $\theta(a) = 0$, so that $z(a) = z_a$.
2. $\theta(T) = 0$ because $z(T)$ is a maximum (or a minimum),
   i.e., $\overline{z}(T) - z(T) \leq 0 \quad (\overline{z}(T) - z(T) \geq 0)$.
3. $\theta(t) = \dfrac{d}{d\epsilon} z(\overline{x}, t)\Big|_{\epsilon=0}$, so that the variation satisfies the differential equation (4.69).

Differentiating θ with respect to t, we obtain that

$$\begin{aligned}
\frac{d}{dt}\theta(t) &= \frac{d}{dt}\frac{d}{d\epsilon} z(\overline{x}, t)\Big|_{\epsilon=0} \\
&= \frac{d}{d\epsilon}\frac{d}{dt} z(\overline{x}, t)\Big|_{\epsilon=0} \\
&= \frac{d}{d\epsilon} L\left(t, x(t) + \epsilon h(t), {}^C D_\gamma^{\alpha(\cdot,\cdot),\beta(\cdot,\cdot)} x(t) + \epsilon\,{}^C D_\gamma^{\alpha(\cdot,\cdot),\beta(\cdot,\cdot)} h(t), \overline{z}(\overline{x}, t)\right)\Big|_{\epsilon=0},
\end{aligned}$$

and rewriting this relation, we obtain the following differential equation for θ:

$$\dot{\theta}(t) - \partial_4 L[x, z]_\gamma^{\alpha,\beta}(t)\theta(t) = \partial_2 L[x, z]_\gamma^{\alpha,\beta}(t)h(t) + \partial_3 L[x, z]_\gamma^{\alpha,\beta}(t)\,{}^C D_\gamma^{\alpha(\cdot,\cdot),\beta(\cdot,\cdot)} h(t).$$

Considering $\lambda(t) = \exp\left(-\int_a^t \partial_4 L[x, z]_\gamma^{\alpha,\beta}(\tau)d\tau\right)$, we obtain the solution for the last differential equation

$$\theta(T)\lambda(T) - \theta(a)$$
$$= \int_a^T \left(\partial_2 L[x, z]_\gamma^{\alpha,\beta}(t)h(t) + \partial_3 L[x, z]_\gamma^{\alpha,\beta}(t)^C D_\gamma^{\alpha(\cdot,\cdot),\beta(\cdot,\cdot)}h(t)\right)\lambda(t)dt.$$

By hypothesis, $\theta(a) = 0$. If x is such that $z(x, t)$ defined by (4.69) attains an extremum at $t = T$, then $\theta(T)$ is identically zero. Hence, we get

$$\int_a^T \left(\partial_2 L[x, z]_\gamma^{\alpha,\beta}(t)h(t) + \partial_3 L[x, z]_\gamma^{\alpha,\beta}(t)^C D_\gamma^{\alpha(\cdot,\cdot),\beta(\cdot,\cdot)}h(t)\right)\lambda(t)dt = 0. \quad (4.74)$$

Considering only the second term in Eq. (4.74) and the definition of combined Caputo derivative operator, we obtain that

$$\int_a^T \lambda(t)\partial_3 L[x, z]_\gamma^{\alpha,\beta}(t)\left(\gamma_1{}_a^C D_t^{\alpha(\cdot,\cdot)}h(t) + \gamma_2{}_t^C D_b^{\beta(\cdot,\cdot)}h(t)\right)dt$$
$$= \gamma_1 \int_a^T \lambda(t)\partial_3 L[x, z]_\gamma^{\alpha,\beta}(t)_a^C D_t^{\alpha(\cdot,\cdot)}h(t)dt$$
$$+ \gamma_2\left[\int_a^b \lambda(t)\partial_3 L[x, z]_\gamma^{\alpha,\beta}(t)_t^C D_b^{\beta(\cdot,\cdot)}h(t)dt \right.$$
$$\left. - \int_T^b \lambda(t)\partial_3 L[x, z]_\gamma^{\alpha,\beta}(t)_t^C D_b^{\beta(\cdot,\cdot)}h(t)dt\right] = \star.$$

Using the fractional integration by parts formula and considering $\bar{\gamma} = (\gamma_2, \gamma_1)$, we get

$$\star = \int_a^T h(t)D_{\bar{\gamma}}^{\beta(\cdot,\cdot),\alpha(\cdot,\cdot)}\left(\lambda(t)\partial_3 L[x, z]_\gamma^{\alpha,\beta}(t)\right)dt$$
$$+ \int_T^b \gamma_2 h(t)\left[_a D_t^{\beta(\cdot,\cdot)}\left(\lambda(t)\partial_3 L[x, z]_\gamma^{\alpha,\beta}(t)\right) -_T D_t^{\beta(\cdot,\cdot)}\left(\lambda(t)\partial_3 L[x, z]_\gamma^{\alpha,\beta}(t)\right)\right]dt$$
$$+ h(T)\left[\gamma_1{}_t I_T^{1-\alpha(\cdot,\cdot)}\left(\lambda(t)\partial_3 L[x, z]_\gamma^{\alpha,\beta}(t)\right) - \gamma_2{}_T I_t^{1-\beta(\cdot,\cdot)}\left(\lambda(t)\partial_3 L[x, z]_\gamma^{\alpha,\beta}(t)\right)\right]_{t=T}$$
$$+ h(b)\gamma_2\left[_T I_t^{1-\beta(\cdot,\cdot)}\left(\lambda(t)\partial_3 L[x, z]_\gamma^{\alpha,\beta}(t)\right) -_a I_t^{1-\beta(\cdot,\cdot)}\left(\lambda(t)\partial_3 L[x, z]_\gamma^{\alpha,\beta}(t)\right)\right]_{t=b}.$$

Substituting this relation into expression (4.74), we obtain

$$0 = \int_a^T h(t) \left[\partial_2 L[x, z]_\gamma^{\alpha,\beta}(t)\lambda(t) + D_{\overline{\gamma}}^{\beta(\cdot,\cdot),\alpha(\cdot,\cdot)} \left(\lambda(t)\partial_3 L[x, z]_\gamma^{\alpha,\beta}(t) \right) \right] dt$$

$$+ \int_T^b \gamma_2 h(t) \left[{}_a D_t^{\beta(\cdot,\cdot)} \left(\lambda(t)\partial_3 L[x, z]_\gamma^{\alpha,\beta}(t) \right) - {}_T D_t^{\beta(\cdot,\cdot)} \left(\lambda(t)\partial_3 L[x, z]_\gamma^{\alpha,\beta}(t) \right) \right] dt$$

$$+ h(T) \left[\gamma_{1t} I_T^{1-\alpha(\cdot,\cdot)} \left(\lambda(t)\partial_3 L[x, z]_\gamma^{\alpha,\beta}(t) \right) - \gamma_{2T} I_t^{1-\beta(\cdot,\cdot)} \left(\lambda(t)\partial_3 L[x, z]_\gamma^{\alpha,\beta}(t) \right) \right]_{t=T}$$

$$+ h(b)\gamma_2 \left[{}_T I_t^{1-\beta(\cdot,\cdot)} \left(\lambda(t)\partial_3 L[x, z]_\gamma^{\alpha,\beta}(t) \right) - {}_a I_t^{1-\beta(\cdot,\cdot)} \left(\lambda(t)\partial_3 L[x, z]_\gamma^{\alpha,\beta}(t) \right) \right]_{t=b}.$$

With appropriate choices for the variations $h(\cdot)$, we get the Euler–Lagrange equations (4.71)–(4.72) and the transversality conditions (4.73).

Remark 58 If $\alpha(\cdot, \cdot)$ and $\beta(\cdot, \cdot)$ tend to 1, and if the Lagrangian L is of class C^2, then the first Euler–Lagrange equation (4.71) becomes

$$\partial_2 L[x, z]_\gamma^{\alpha,\beta}(t)\lambda(t) + (\gamma_2 - \gamma_1)\frac{d}{dt}\left[\lambda(t)\partial_3 L[x, z]_\gamma^{\alpha,\beta}(t) \right] = 0.$$

Differentiating and considering the derivative of the lambda function, we obtain

$$\lambda(t)\left[\partial_2 L[x, z]_\gamma^{\alpha,\beta}(t) \right.$$

$$\left. + (\gamma_2 - \gamma_1)\left[-\partial_4 L[x, z]_\gamma^{\alpha,\beta}(t)\partial_3 L[x, z]_\gamma^{\alpha,\beta}(t) + \frac{d}{dt}\partial_3 L[x, z]_\gamma^{\alpha,\beta}(t) \right] \right] = 0.$$

As $\lambda(t) > 0$, for all t, we deduce that

$$\partial_2 L[x, z]_\gamma^{\alpha,\beta}(t) + (\gamma_2 - \gamma_1)\left[\frac{d}{dt}\partial_3 L[x, z]_\gamma^{\alpha,\beta}(t) - \partial_4 L[x, z]_\gamma^{\alpha,\beta}(t)\partial_3 L[x, z]_\gamma^{\alpha,\beta}(t) \right] = 0.$$

4.7.2 Several Independent Variables

Consider the following generalization of the problem of Herglotz involving $n + 1$ independent variables. Define $\Omega = \prod_{i=1}^n [a_i, b_i]$, with $n \in \mathbb{N}$, $P = [a, b] \times \Omega$, and consider the vector $s = (s_1, s_2, \ldots, s_n) \in \Omega$. The new problem consists in determining the trajectories $x \in C^1(P)$ that give an extremum to $z[x, T]$, where the functional z satisfies the differential equation

$$\dot{z}(t) = \int_\Omega L\left(t, s, x(t, s), {}^C D_\gamma^{\alpha(\cdot,\cdot),\beta(\cdot,\cdot)} x(t, s), \right.$$

$$\left. {}^C D_{\gamma^1}^{\alpha_1(\cdot,\cdot),\beta_1(\cdot,\cdot)} x(t, s), \ldots, {}^C D_{\gamma^n}^{\alpha_n(\cdot,\cdot),\beta_n(\cdot,\cdot)} x(t, s), z(t) \right) d^n s \qquad (4.75)$$

subject to the constraint

$$x(t, s) = g(t, s), \quad \text{for all} \quad (t, s) \in \partial P, \tag{4.76}$$

where ∂P is the boundary of P and g is a given function $g : \partial P \to \mathbb{R}$. We assume that

1. $\alpha, \alpha_i, \beta, \beta_i : [a, b]^2 \to (0, 1)$ with $i = 1, \ldots, n$.
2. $\gamma, \gamma^1, \ldots, \gamma^n \in [0, 1]^2$.
3. $d^n s = ds_1 \ldots ds_n$.
4. ${}^C D_\gamma^{\alpha(\cdot,\cdot),\beta(\cdot,\cdot)} x(t, s)$, ${}^C D_{\gamma^1}^{\alpha_1(\cdot,\cdot),\beta_1(\cdot,\cdot)} x(t, s), \ldots, {}^C D_{\gamma^n}^{\alpha_n(\cdot,\cdot),\beta_n(\cdot,\cdot)} x(t, s)$ exist and are continuous functions.
5. The Lagrangian $L : P \times \mathbb{R}^{n+3} \to \mathbb{R}$ is of class C^1.

Remark 59 By ${}^C D_\gamma^{\alpha(\cdot,\cdot),\beta(\cdot,\cdot)} x(t, s)$, we mean the Caputo fractional derivative with respect to the independent variable t, and by ${}^C D_{\gamma^i}^{\alpha_i(\cdot,\cdot),\beta_i(\cdot,\cdot)} x(t, s)$ we mean the Caputo fractional derivative with respect to the independent variable s_i, for $i = 1, \ldots, n$.

In the sequel, we use the auxiliary notation $[x, z]_{n,\gamma}^{\alpha,\beta}(t, s)$ to represent the following vector

$$\left(t, s, x(t, s), {}^C D_\gamma^{\alpha(\cdot,\cdot),\beta(\cdot,\cdot)} x(t, s), {}^C D_{\gamma^1}^{\alpha_1(\cdot,\cdot),\beta_1(\cdot,\cdot)} x(t, s), \right.$$

$$\left. \ldots, {}^C D_{\gamma^n}^{\alpha_n(\cdot,\cdot),\beta_n(\cdot,\cdot)} x(t, s), z(t) \right).$$

Consider the function

$$\lambda(t) = \exp\left(-\int_a^t \int_\Omega \partial_{2n+4} [x, z]_{n,\gamma}^{\alpha,\beta}(\tau, s) d^n s d\tau \right).$$

Theorem 60 *If (x, z, T) is an extremizer of the functional defined by Eq. (4.75), then (x, z, T) satisfies the fractional differential equations*

$$\partial_{n+2} L[x, z]_{n,\gamma}^{\alpha,\beta}(t, s) \lambda(t) + D_{\overline{\gamma}}^{\beta(\cdot,\cdot),\alpha(\cdot,\cdot)} \left(\lambda(t) \partial_{n+3} L[x, z]_{n,\gamma}^{\alpha,\beta}(t, s) \right)$$

$$+ \sum_{i=1}^n D_{\overline{\gamma^i}}^{\beta_i(\cdot,\cdot),\alpha_i(\cdot,\cdot)} \left(\lambda(t) \partial_{n+3+i} L[x, z]_{n,\gamma}^{\alpha,\beta}(t, s) \right) = 0 \tag{4.77}$$

on $[a, T] \times \Omega$ and

$$\gamma_2 \left({}_a D_t^{\beta(\cdot,\cdot)} \left(\lambda(t) \partial_{n+3} L[x, z]_{n,\gamma}^{\alpha,\beta}(t, s) \right) - {}_T D_t^{\beta(\cdot,\cdot)} \left(\lambda(t) \partial_{n+3} L[x, z]_{n,\gamma}^{\alpha,\beta}(t, s) \right) \right) = 0 \tag{4.78}$$

on $[T, b] \times \Omega$.

Moreover, (x, z) satisfies the transversality condition

$$\left[\gamma_{1t} I_T^{1-\alpha(\cdot,\cdot)} \left(\lambda(t) \partial_{n+3} L[x, z]_{n,\gamma}^{\alpha,\beta}(t, s) \right) \right.$$

$$\left. - \gamma_{2T} I_t^{1-\beta(\cdot,\cdot)} \left(\lambda(t) \partial_{n+3} L[x, z]_{n,\gamma}^{\alpha,\beta}(t, s) \right) \right]_{t=T} = 0, \quad s \in \Omega. \tag{4.79}$$

If $T < b$, then $\int_\Omega L[x, z]_{n,\gamma}^{\alpha,\beta}(T, s)d^n s = 0$.

Proof Let x be a solution to the problem. Consider an admissible variation of x, $\overline{x}(t, s) = x(t, s) + \epsilon h(t, s)$, where $h \in C^1(P)$ is an arbitrary perturbing curve and $\epsilon \in \mathbb{R}$ is such that $|\epsilon| \ll 1$. Consequently, $h(t, s) = 0$ for all $(t, s) \in \partial P$ by the boundary condition (4.76).

On the other hand, consider an admissible variation of z, $\overline{z} = z + \epsilon\theta$, where θ is a perturbing curve such that $\theta(a) = 0$ and

$$\theta(t) = \frac{d}{d\varepsilon} z(\overline{x}, t)\bigg|_{\varepsilon=0}.$$

Differentiating $\theta(t)$ with respect to t, we obtain that

$$\frac{d}{dt}\theta(t) = \frac{d}{dt}\frac{d}{d\varepsilon} z(\overline{x}, t)\bigg|_{\varepsilon=0}$$

$$= \frac{d}{d\varepsilon}\frac{d}{dt} z(\overline{x}, t)\bigg|_{\varepsilon=0}$$

$$= \frac{d}{d\varepsilon}\int_\Omega L[\overline{x}, z]_{n,\gamma}^{\alpha,\beta}(t, s)d^n s\bigg|_{\varepsilon=0}.$$

We conclude that

$$\dot{\theta}(t) = \int_\Omega \left(\partial_{n+2}L[x, z]_{n,\gamma}^{\alpha,\beta}(t, s)h(t, s) + \partial_{n+3}L[x, z]_{n,\gamma}^{\alpha,\beta}(t, s)^C D_\gamma^{\alpha(\cdot,\cdot),\beta(\cdot,\cdot)}h(t, s)\right.$$

$$+ \sum_{i=1}^n \partial_{n+3+i}L[x, z]_{n,\gamma}^{\alpha,\beta}(t, s)^C D_{\gamma^i}^{\alpha_i(\cdot,\cdot),\beta_i(\cdot,\cdot)}h(t, s) + \partial_{2n+4}L[x, z]_{n,\gamma}^{\alpha,\beta}(t, s)\theta(t)\bigg) d^n s.$$

To simplify the notation, we define

$$B(t) = \int_\Omega \partial_{2n+4}L[x, z]_{n,\gamma}^{\alpha,\beta}(t, s)d^n s$$

and

$$A(t) = \int_\Omega \left(\partial_{n+2}L[x, z]_{n,\gamma}^{\alpha,\beta}(t, s)h(t, s) + \partial_{n+3}L[x, z]_{n,\gamma}^{\alpha,\beta}(t, s)^C D_\gamma^{\alpha(\cdot,\cdot),\beta(\cdot,\cdot)}h(t, s)\right.$$

$$+ \sum_{i=1}^n \partial_{n+3+i}L[x, z]_{n,\gamma}^{\alpha,\beta}(t, s)^C D_{\gamma^i}^{\alpha_i(\cdot,\cdot),\beta_i(\cdot,\cdot)}h(t, s)\bigg) d^n s.$$

Then, we obtain the linear differential equation

$$\dot{\theta}(t) - B(t)\theta(t) = A(t),$$

whose solution is

$$\theta(T)\lambda(T) - \theta(a) = \int_a^T A(t)\lambda(t)dt.$$

Since $\theta(a) = \theta(T) = 0$, we get

$$\int_a^T A(t)\lambda(t)dt = 0. \qquad (4.80)$$

Considering only the second term in (4.80), we can write

$$\int_a^T \int_\Omega \lambda(t)\partial_{n+3}L[x, z]_{n,\gamma}^{\alpha,\beta}(t, s) \left(\gamma_1{}_a^C D_t^{\alpha(\cdot,\cdot)}h(t, s) + \gamma_2{}_t^C D_b^{\beta(\cdot,\cdot)}h(t, s)\right) d^n s dt$$

$$= \gamma_1 \int_a^T \int_\Omega \lambda(t)\partial_{n+3}L[x, z]_{n,\gamma}^{\alpha,\beta}(t, s)_a^C D_t^{\alpha(\cdot,\cdot)}h(t, s)d^n s dt$$

$$+ \gamma_2 \left[\int_a^b \int_\Omega \lambda(t)\partial_{n+3}L[x, z]_{n,\gamma}^{\alpha,\beta}(t, s)_t^C D_b^{\beta(\cdot,\cdot)}h(t, s)d^n s dt \right.$$

$$\left. - \int_T^b \int_\Omega \lambda(t)\partial_{n+3}L[x, z]_{n,\gamma}^{\alpha,\beta}(t, s)_t^C D_b^{\beta(\cdot,\cdot)}h(t, s)d^n s dt\right].$$

Let $\overline{\gamma} = (\gamma_2, \gamma_1)$. Integrating by parts (cf. Theorem 13), and since $h(a, s) = h(b, s) = 0$ for all $s \in \Omega$, we obtain the following expression:

$$\int_a^T \int_\Omega h(t, s) D_{\overline{\gamma}}^{\beta(\cdot,\cdot),\alpha(\cdot,\cdot)} \left(\lambda(t)\partial_{n+3}L[x, z]_{n,\gamma}^{\alpha,\beta}(t, s)\right) d^n s dt$$

$$+ \gamma_2 \int_T^b \int_\Omega h(t, s) \left[_a D_t^{\beta(\cdot,\cdot)} \left(\lambda(t)\partial_{n+3}L[x, z]_{n,\gamma}^{\alpha,\beta}(t, s)\right)\right.$$

$$\left. - _T D_t^{\beta(\cdot,\cdot)} \left(\lambda(t)\partial_{n+3}L[x, z]_{n,\gamma}^{\alpha,\beta}(t, s)\right)\right] d^n s dt$$

$$+ \int_\Omega h(T, s) \left[\gamma_1{}_t I_T^{1-\alpha(\cdot,\cdot)} \left(\lambda(t)\partial_{n+3}L[x, z]_{n,\gamma}^{\alpha,\beta}(t, s)\right)\right.$$

$$\left. - \gamma_2{}_T I_t^{1-\beta(\cdot,\cdot)} \left(\lambda(t)\partial_{n+3}L[x, z]_{n,\gamma}^{\alpha,\beta}(t, s)\right) d^n s\right]_{t=T}.$$

Doing similarly for the $(i + 2)$th term in (4.80), with $i = 1, \ldots, n$, letting $\overline{\gamma}^i = (\gamma_2^i, \gamma_1^i)$, and since $h(t, a_i) = h(t, b_i) = 0$ for all $t \in [a, b]$, we obtain

$$\int_a^T \int_\Omega \lambda(t)\partial_{n+3+i}L[x, z]_{n,\gamma}^{\alpha,\beta}(t, s) \left(\gamma_{1a_i}^{iC} D_{s_i}^{\alpha_i(\cdot,\cdot)}h(t, s) + \gamma_{2s_i}^{iC} D_{b_i}^{\beta_i(\cdot,\cdot)}h(t, s)\right) d^n s dt$$

$$= \int_a^T \int_\Omega h(t, s) D_{\overline{\gamma}^i}^{\beta_i(\cdot,\cdot),\alpha_i(\cdot,\cdot)} \left(\lambda(t)\partial_{n+3+i}L[x, z]_{n,\gamma}^{\alpha,\beta}(t, s)\right) d^n s dt.$$

Substituting these relations into (4.80), we deduce that

$$\int_a^T \int_\Omega h(t,s) \Big[\partial_{n+2}L[x,z]_{n,\gamma}^{\alpha,\beta}(t,s)\lambda(t) + D_{\overline{\gamma}}^{\beta(\cdot,\cdot),\alpha(\cdot,\cdot)} \left(\lambda(t)\partial_{n+3}L[x,z]_{n,\gamma}^{\alpha,\beta}(t,s) \right)$$

$$+ \sum_{i=1}^n D_{\overline{\gamma}^i}^{\beta_i(\cdot,\cdot),\alpha_i(\cdot,\cdot)} \left(\lambda(t)\partial_{n+3+i}L[x,z]_{n,\gamma}^{\alpha,\beta}(t,s) \right) d^n s dt$$

$$+ \gamma_2 \int_T^b \int_\Omega h(t,s) \Big[{}_aD_t^{\beta(\cdot,\cdot)} \left(\lambda(t)\partial_{n+3}L[x,z]_{n,\gamma}^{\alpha,\beta}(t,s) \right)$$

$$-_T D_t^{\beta(\cdot,\cdot)} \left(\lambda(t)\partial_{n+3}L[x,z]_{n,\gamma}^{\alpha,\beta}(t,s) \right) \Big] d^n s dt$$

$$+ \int_\Omega h(T,s) \Big[\gamma_{1t} I_T^{1-\alpha(\cdot,\cdot)} \left(\lambda(t)\partial_{n+3}L[x,z]_{n,\gamma}^{\alpha,\beta}(t,s) \right)$$

$$-\gamma_{2\,T} I_t^{1-\beta(\cdot,\cdot)} \left(\lambda(t)\partial_{n+3}L[x,z]_{n,\gamma}^{\alpha,\beta}(t,s) \right) d^n s \Big]_{t=T}.$$

For appropriate choices with respect to h, we get the Euler–Lagrange equations (4.77)–(4.78) and the transversality condition (4.79).

4.7.3 Examples

We present three examples, with and without the dependence on z.

Example 4.1 Consider

$$\dot{z}(t) = \left({}^C D_\gamma^{\alpha(\cdot,\cdot),\beta(\cdot,\cdot)} x(t) \right)^2 + z(t) + t^2 - 1, \quad t \in [0,3], \tag{4.81}$$
$$x(0) = 1, \quad z(0) = 0.$$

In this case, $\lambda(t) = \exp(-t)$. The necessary optimality conditions (4.71)–(4.72) of Theorem 57 hold for $\overline{x}(t) \equiv 1$. If we replace x by $\overline{x}$ in (4.81), we obtain

$$\dot{z}(t) - z(t) = t^2 - 1, \quad t \in [0,3],$$
$$z(0) = 0,$$

whose solution is

$$z(t) = \exp(t) - (t+1)^2. \tag{4.82}$$

The last transversality condition of Theorem 57 asserts that

$$L[\overline{x},z]_\gamma^{\alpha,\beta}(T) = 0 \Leftrightarrow \exp(T) - 2T - 2 = 0,$$

whose solution is approximately

$$T \approx 1.67835.$$

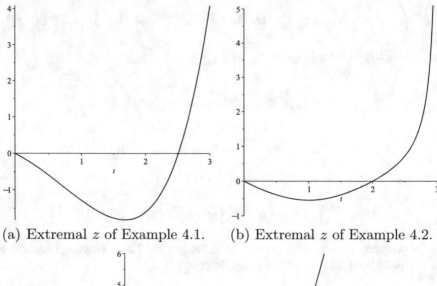

(a) Extremal z of Example 4.1.   (b) Extremal z of Example 4.2.

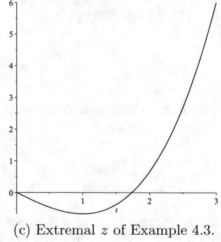

(c) Extremal z of Example 4.3.

Fig. 4.1 Graphics of function $z(\overline{x}, t)$

We remark that z (4.82) actually attains a minimum value at this point (see Fig. 4.1a):

$$z(1.67835) \approx -1.81685.$$

Example 4.2 Consider now

$$\dot{z}(t) = (t - 1)\left(x^2(t) + z^2(t) + 1\right), \quad t \in [0, 3],$$
$$x(0) = 0, \quad z(0) = 0.$$

(4.83)

Since the first Euler–Lagrange equation (4.71) reads

$$(t-1)x(t) = 0, \quad \forall t \in [0, T],$$

we see that $\bar{x}(t) \equiv 0$ is a solution to this equation. The second transversality condition of (4.73) asserts that, at $t = T$, we must have

$$L[\bar{x}, z]_\gamma^{\alpha,\beta}(t) = 0,$$

that is,

$$(t-1)(z^2(t) + 1) = 0,$$

and so $T = 1$ is a solution to this equation. Substituting x by $\bar{x}$ in (4.83), we get

$$\dot{z}(t) = (t-1)(z^2(t) + 1), \quad t \in [0, 3],$$
$$z(0) = 0.$$

The solution to this Cauchy problem is the function

$$z(t) = \tan\left(\frac{t^2}{2} - t\right),$$

(see Fig. 4.1b) and the minimum value is

$$z(1) = \tan\left(-\frac{1}{2}\right).$$

Example 4.3 For our last example, consider

$$\dot{z}(t) = \left({}^C D_\gamma^{\alpha(\cdot,\cdot),\beta(\cdot,\cdot)} x(t) - f(t)\right)^2 + t^2 - 1, \quad t \in [0, 3], \tag{4.84}$$
$$x(0) = 0, \quad z(0) = 0,$$

where

$$f(t) := \frac{t^{1-\alpha(t)}}{2\Gamma(2 - \alpha(t))} - \frac{(3-t)^{1-\beta(t)}}{2\Gamma(2 - \beta(t))}.$$

In this case, $\lambda(t) \equiv 1$. We intend to find a pair (x, z), satisfying all the conditions in (4.84), for which $z(T)$ attains a minimum value. It is easy to verify that $\bar{x}(t) = t$ and $T = 1$ satisfy the necessary conditions given by Theorem 57. Replacing x by $\bar{x}$ in (4.84), we get a Cauchy problem of form

$$\dot{z}(t) = t^2 - 1, \quad t \in [0, 3],$$
$$z(0) = 0,$$

whose solution is

$$z(t) = \frac{t^3}{3} - t.$$

Observe that this function attains a minimum value at $T = 1$, which is $z(1) = -2/3$ (Fig. 4.1c).

References

1. Almeida R (2016) Fractional variational problems depending on indefinite integrals and with delay. Bull Malay Math Sci Soc 39(4):1515–1528
2. Almeida R, Malinowska AB (2014) Fractional variational principle of Herglotz. Discret Contin Dyn Syst Ser B 19(8):2367–2381
3. Baleanu D, Maaraba T, Jarad F (2008) Fractional variational principles with delay. J Phys A 41(31):315403 8 pp
4. Caputo MC, Torres DFM (2015) Duality for the left and right fractional derivatives. Signal Process 107:265–271
5. Daftardar-Gejji V, Sukale Y, Bhalekar S (2015) Solving fractional delay differential equations: A new approach. Fract Calc Appl Anal 16:400–418
6. Deng W, Li C, Lü J (2007) Stability analysis of linear fractional differential system with multiple time delays. Nonlinear Dyn 48(4):409–416
7. Georgieva B, Guenther RB (2002) First Noether-type theorem for the generalized variational principle of Herglotz. Topol Methods Nonlinear Anal 20(2):261–273
8. Georgieva B, Guenther RB, Bodurov T (2003) Generalized variational principle of Herglotz for several independent variables. J Math Phys 44(9):3911–3927
9. Guenther RB, Gottsch JA, Kramer DB (1996) The Herglotz algorithm for constructing canonical transformations. SIAM Rev 38(2):287–293
10. Guenther RB, Guenther CM, Gottsch JA (1996) The Herglotz lectures on contact transformations and hamiltonian systems, vol 1. Lecture notes in nonlinear analysis. Juliusz Schauder Center for Nonlinear Studies, Nicholas Copernicus University, Torún
11. Herglotz G (1930) Berührungstransformationen. Lectures at the University of Göttingen, Göttingen
12. Jarad F, Abdeljawad T, Baleanu D (2010) Fractional variational principles with delay within Caputo derivatives. Rep Math Phys 65(1):17–28
13. Lazarević MP, Spasić AM (2009) Finite-time stability analysis of fractional order time-delay systems: Gronwall's approach. Math Comput Model 49(3–4):475–481
14. Linge S, Langtangen HP (2016) Programming for computations–MATLAB/Octave. A gentle introduction to numerical simulations with MATLAB/Octave. Springer, Cham
15. Machado JAT (2011) Time-delay and fractional derivatives. Adv Differ Equ 2011:934094. 12 pp
16. Malinowska AB, Odzijewicz T, Torres DFM (2015) Advanced methods in the fractional calculus of variations. Springer briefs in applied sciences and technology. Springer, Cham
17. Santos SPS, Martins N, Torres DFM (2014) Higher-order variational problems of Herglotz type. Vietnam J Math 42(4):409–419
18. Santos SPS, Martins N, Torres DFM (2015) Variational problems of Herglotz type with time delay: DuBois-Reymond condition and Noether's first theorem. Discret Contin Dyn Syst 35(9):4593–4610
19. Santos SPS, Martins N, Torres DFM (2015) An optimal control approach to Herglotz variational problems. In: Plakhov A, Tchemisova T, Freitas A (eds) Optimization in the natural sciences, Communications in computer and information science, Vol. 499. Springer, pp 107–117

20. Tavares D, Almeida R, Torres DFM (2015) Optimality conditions for fractional variational problems with dependence on a combined Caputo derivative of variable order. Optimization 64(6):1381–1391
21. Tavares D, Almeida R, Torres DFM (2017) Constrained fractional variational problems of variable order. IEEE/CAA Jl Autom Sinica 4(1):80–88
22. Tavares D, Almeida R, Torres DFM (2018) Fractional Herglotz variational problem of variable order. Discret Contin Dyn Syst Ser S 11(1):143–154
23. Tavares D, Almeida R, Torres DFM (2018) Combined fractional variational problems of variable order and some computational aspects. J Comput Appl Math 339:374–388
24. Trefethen LN (2013) Approximation theory and approximation practice. Society for industrial and applied mathematics
25. van Brunt B (2004) The calculus of variations. Universitext. Springer, New York
26. Wang H, Yu Y, Wen G, Zhang S (2015) Stability analysis of fractional order neural networks with time delay. Neural Process Lett 42(2):479–500

Appendix

In this appendix, we use a specific **MATLAB** software, the package **Chebfun**, to obtain a few computational approximations for the main fractional operators in this book.

Chebfun is an open-source software package that "aims to provide numerical computing with functions" in **MATLAB** [2]. **Chebfun** overloads **MATLAB**'s discrete operations for matrices to analogous continuous operations for functions and operators [3]. For the mathematical underpinnings of **Chebfun**, we refer the reader to Trefethen [3]. For the algorithmic backstory of **Chebfun**, we refer to Driscoll et al. [1].

In what follows, we study some computational approximations of Riemann–Liouville fractional integrals, of Caputo fractional derivatives and consequently of the combined Caputo fractional derivative, all of them with variable order. We provide, also, the necessary **Chebfun** code for the variable-order fractional calculus.

To implement these operators, we need two auxiliary functions: the gamma function Γ (Definition 1) and the beta function B (Definition 3). Both functions are available in **MATLAB** through the commands `gamma(t)` and `beta(t,u)`, respectively.

A.1 Higher-Order Riemann–Liouville Fractional Integrals

In this section, we discuss computational aspects to the higher-order Riemann–Liouville fractional integrals of variable-order $_aI_t^{\alpha_n(\cdot,\cdot)}x(t)$ and $_tI_b^{\alpha_n(\cdot,\cdot)}x(t)$.

Considering the Definition 34 of higher-order Riemann–Liouville fractional integrals, we implemented in **Chebfun** two functions `leftFi(x,alpha,a)` and `rightFI(x,alpha,b)` that approximate, respectively, the Riemann–Liouville fractional integrals $_aI_t^{\alpha_n(\cdot,\cdot)}x(t)$ and $_tI_b^{\alpha_n(\cdot,\cdot)}x(t)$, through the following **Chebfun/ MATLAB** code.

© The Author(s), under exclusive license to Springer International
Publishing AG, part of Springer Nature 2019
R. Almeida et al., *The Variable-Order Fractional Calculus of Variations*, SpringerBriefs
in Applied Sciences and Technology, https://doi.org/10.1007/978-3-319-94006-9

```
function r = leftFI(x,alpha,a)
g = @(t,tau) x(tau)./(gamma(alpha(t,tau)).*(t-tau).^(1-alpha(t,tau)));
r = @(t) sum(chebfun(@(tau) g(t,tau),[a t],'splitting','on'),[a t]);
end
```

and

```
function r = rightFI(x,alpha,b)
g = @(t,tau) x(tau)./(gamma(alpha(tau,t)).*(tau-t).^(1-alpha(tau,t)));
r = @(t) sum(chebfun(@(tau) g(t,tau),[t b],'splitting','on'),[t b]);
end
```

With these two functions, we illustrate their use in the following example, where we determine computacional approximations for Riemann–Liouville fractional integrals of a special power function.

Example 4.4 Let $\alpha(t, \tau) = \frac{t^2+\tau^2}{4}$ and $x(t) = t^2$ with $t \in [0, 1]$. In this case, $a = 0$, $b = 1$ and $n = 1$. We have $_aI_{0.6}^{\alpha(\cdot,\cdot)}x(0.6) \approx 0.2661$ and $_{0.6}I_b^{\alpha(\cdot,\cdot)}x(0.6) \approx 0.4619$, obtained in MATLAB with our Chebfun functions as follows:

```
a = 0; b = 1; n = 1;
alpha = @(t,tau) (t.^2+tau.^2)/4;
x = chebfun(@(t) t.^2, [0,1]);
LFI = leftFI(x,alpha,a);
RFI = rightFI(x,alpha,b);
LFI(0.6)
ans = 0.2661
RFI(0.6)
ans = 0.4619
```

Other values for $_aI_t^{\alpha(\cdot,\cdot)}x(t)$ and $_tI_b^{\alpha(\cdot,\cdot)}x(t)$ are plotted in Fig. A.1.

A.2   Higher-Order Caputo Fractional Derivatives

In this section, considering the Definition 36, we implement in Chebfun two new functions leftCaputo(x,alpha,a,n) and rightCaputo(x,alpha, b,n) that approximate, respectively, the higher-order Caputo fractional derivatives of variable-order $_a^CD_t^{\alpha_n(\cdot,\cdot)}x(t)$ and $_t^CD_b^{\alpha_n(\cdot,\cdot)}x(t)$.

The following code implements the left operator (4.2):

```
function r = leftCaputo(x,alpha,a,n)
dx = diff(x,n);
g = @(t,tau) dx(tau)./(gamma(n-alpha(t,tau)).
                    *(t-tau).^(1+alpha(t,tau)-n));
r = @(t) sum(chebfun(@(tau) g(t,tau),[a t],'splitting','on'),[a t]);
end
```

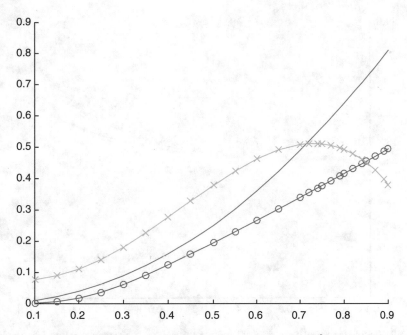

Fig. A.1 Riemann–Liouville fractional integrals of Example 4.4: $x(t) = t^2$ in continuous line, left integral $_aI_t^{\alpha(\cdot,\cdot)}x(t)$ with "o−" style, and right integral $_tI_b^{\alpha(\cdot,\cdot)}x(t)$ with "×−" style

Similarly, we define the right operator (4.3) with **Chebfun** in **MATLAB** as follows:

```
function r = rightCaputo(x,alpha,b,n)
dx = diff(x,n);
g = @(t,tau) dx(tau)./(gamma(n-alpha(tau,t)).
                    *(tau-t).^(1+alpha(tau,t)-n));
r = @(t)(-1).^n.* sum(chebfun(@(tau) g(t,tau),[t b],
                    'splitting','on'),[t b]);
end
```

We use the two functions `leftCaputo` and `rightCaputo` to determine aproximations for the Caputo fractional derivatives of a power function of the form $x(t) = t^\gamma$.

Example 4.5 Let $\alpha(t,\tau) = \frac{t^2}{2}$ and $x(t) = t^4$ with $t \in [0, 1]$. In this case, $a = 0$, $b = 1$ and $n = 1$. We have $_a^C D_{0.6}^{\alpha(\cdot,\cdot)}x(0.6) \approx 0.1857$ and $_{0.6}^C D_b^{\alpha(\cdot,\cdot)}x(0.6) \approx -1.0385$, obtained in **MATLAB** with our **Chebfun** functions as follows:

```
a = 0; b = 1; n = 1;
alpha = @(t,tau) t.^2/2;
x = chebfun(@(t) t.^4, [a b]);
LC = leftCaputo(x,alpha,a,n);
RC = rightCaputo(x,alpha,b,n);
LC(0.6)
```

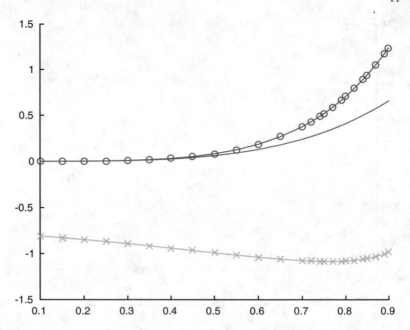

Fig. A.2 Caputo fractional derivatives of Example 4.5: $x(t) = t^4$ in continuous line, left derivative ${}^C_a D_t^{\alpha(\cdot,\cdot)} x(t)$ with "○−" style, and right derivative ${}^C_t D_b^{\alpha(\cdot,\cdot)} x(t)$ with "×−" style

```
ans = 0.1857
RC(0.6)
ans = -1.0385
```

See Fig. A.2 for a plot with other values of ${}^C_a D_t^{\alpha(\cdot,\cdot)} x(t)$ and ${}^C_t D_b^{\alpha(\cdot,\cdot)} x(t)$.

Example 4.6 In Example 4.5, we have used the polynomial $x(t) = t^4$. It is worth mentioning that our **Chebfun** implementation works well for functions that are not a polynomial. For example, let $x(t) = e^t$. In this case, we just need to change

```
x = chebfun(@(t) t.^4, [a b]);
```

in Example 4.5 by

```
x = chebfun(@(t) exp(t), [a b]);
```

to obtain

```
LC(0.6)
ans = 0.9917
RC(0.6)
ans = -1.1398
```

See Fig. A.3 for a plot with other values of ${}^C_a D_t^{\alpha(\cdot,\cdot)} x(t)$ and ${}^C_t D_b^{\alpha(\cdot,\cdot)} x(t)$.

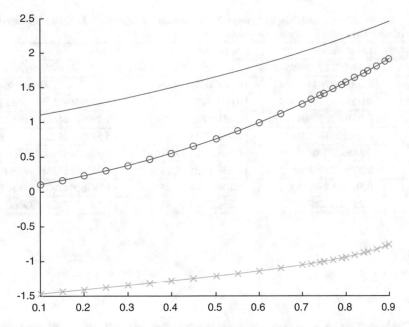

Fig. A.3 Caputo fractional derivatives of Example 4.6: $x(t) = e^t$ in continuous line, left derivative ${}^{C}_{a}D^{\alpha(\cdot,\cdot)}_{t}x(t)$ with "∘−" style, and right derivative ${}^{C}_{t}D^{\alpha(\cdot,\cdot)}_{b}x(t)$ with "×−" style

With Lemma 40 in Sect. 4.1, we can obtain, analytically, the higher-order left Caputo fractional derivative of a power function of the form $x(t) = (t-a)^{\gamma}$. This allows us to show the effectiveness of our computational approach, that is, the usefulness of polynomial interpolation in Chebyshev points in fractional calculus of variable order. In Lemma 40, we assume that the fractional order depends only on the first variable: $\alpha_n(t, \tau) := \overline{\alpha}_n(t)$, where $\overline{\alpha}_n : [a, b] \to (n-1, n)$ is a given function.

Example 4.7 Let us revisit Example 4.5 by choosing $\alpha(t, \tau) = \frac{t^2}{2}$ and $x(t) = t^4$ with $t \in [0, 1]$. Table A.1 shows the approximated values obtained by our **Chebfun** function `leftCaputo(x,alpha,a,n)` and the exact values computed with the formula given by Lemma 40. Table A.1 was obtained using the following MATLAB code:

```
format long
a = 0; b = 1; n = 1;
alpha = @(t,tau) t.^2/2;
x = chebfun(@(t) t.^4, [a b]);
exact = @(t) (gamma(5)./gamma(5-alpha(t))).*t.^(4-alpha(t));
approximation = leftCaputo(x,alpha,a,n);
for i = 1:9
t = 0.1*i;
E = exact(t);
A = approximation(t);
```

Table A.1 Exact values obtained by Lemma 40 for functions of Example 4.7 versus computational approximations obtained using the Chebfun code

t	Exact Value	Approximation	Error
0.1	1.019223177296953e-04	1.019223177296974e-04	−2.046431600566390e-18
0.2	0.001702793965464	0.001702793965464	−2.168404344971009e-18
0.3	0.009148530806348	0.009148530806348	3.469446951953614e-18
0.4	0.031052290994593	0.031052290994592	9.089951014118469e-16
0.5	0.082132144921157	0.082132144921157	6.522560269672795e-16
0.6	0.185651036003120	0.185651036003112	7.938094626069869e-15
0.7	0.376408251363662	0.376408251357416	6.246059225389899e-12
0.8	0.704111480975332	0.704111480816562	1.587694420379648e-10
0.9	1.236753486749357	1.236753486514274	2.350835082154390e-10

```
error = E - A;
[t E A error]
end
```

Computational experiments similar to those of Example 4.7, obtained by substituting Lemma 40 by Lemma 41 and our `leftCaputo` routine by the `rightCaputo` one, reinforce the validity of our computational methods. In this case, we assume that the fractional order depends only on the second variable: $\alpha_n(\tau, t) := \overline{\alpha}_n(t)$, where $\overline{\alpha}_n : [a, b] \to (n - 1, n)$ is a given function.

A.3 Higher-order Combined Fractional Caputo Derivative

The higher-order combined Caputo fractional derivative combines both left and right Caputo fractional derivatives; that is, we make use of functions `leftCaputo` `(x,alpha,a,n)` and `rightCaputo(x,alpha,b,n)` provided in Sect. A.1 to define Chebfun computational code for the higher-order combined fractional Caputo derivative of variable order:

```
function r = combinedCaputo(x,alpha,beta,gamma1,gamma2,a,b,n)
lc = leftCaputo(x,alpha,a,n);
rc = rightCaputo(x,beta,b,n);
r = @(t) gamma1 .* lc(t) + gamma2 .* rc(t);
end
```

Then, we illustrate the behavior of the combined Caputo fractional derivative of variable order for different values of $t \in (0, 1)$, using MATLAB.

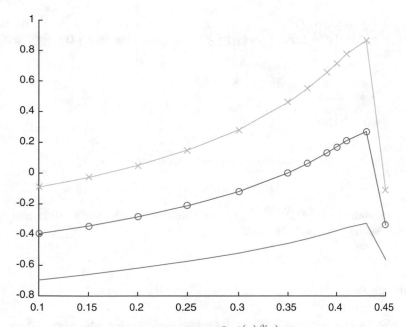

Fig. A.4 Combined Caputo fractional derivative $^{C}D_{\gamma}^{\alpha(\cdot,\cdot),\beta(\cdot,\cdot)}x(t)$ for $\alpha(\cdot,\cdot)$, $\beta(\cdot,\cdot)$ and $x(t)$ of Example 4.8: continuous line for $\gamma = (\gamma_1, \gamma_2) = (0.2, 0.8)$, "$\circ-$" style for $\gamma = (\gamma_1, \gamma_2) = (0.5, 0.5)$, and "$\times-$" style for $\gamma = (\gamma_1, \gamma_2) = (0.8, 0.2)$

Table A.2 Combined Caputo fractional derivative $^{C}D_{\gamma}^{\alpha(\cdot,\cdot),\beta(\cdot,\cdot)}x(t)$ for $\alpha(\cdot,\cdot)$, $\beta(\cdot,\cdot)$ and $x(t)$ of Example 4.8. Case 1: $\gamma = (\gamma_1, \gamma_2) = (0.2, 0.8)$; Case 2: $\gamma = (\gamma_1, \gamma_2) = (0.5, 0.5)$; Case 3: $\gamma = (\gamma_1, \gamma_2) = (0.8, 0.2)$

t	Case 1	Case 2	Case 3
0.4500	−0.5630	−0.3371	−0.1112
0.5000	−790.4972	−1.9752e+03	−3.1599e+03
0.5500	−3.5738e+06	−8.9345e+06	−1.4295e+07
0.6000	−2.0081e+10	−5.0201e+10	−8.0322e+10
0.6500	2.8464e+14	7.1160e+14	1.1386e+15
0.7000	4.8494e+19	1.2124e+20	1.9398e+20
0.7500	3.8006e+24	9.5015e+24	1.5202e+25
0.8000	−1.3648e+30	−3.4119e+30	−5.4591e+30
0.8500	−1.6912e+36	−4.2280e+36	−6.7648e+36
0.9000	5.5578e+41	1.3895e+42	2.2231e+42
0.9500	1.5258e+49	3.8145e+49	6.1033e+49
0.9900	1.8158e+54	4.5394e+54	7.2631e+54

Example 4.8 Let $\alpha(t, \tau) = \frac{t^2 + \tau^2}{4}$, $\beta(t, \tau) = \frac{t + \tau}{3}$ and $x(t) = t$, $t \in [0, 1]$. We have $a = 0$, $b = 1$ and $n = 1$. For $\gamma = (\gamma_1, \gamma_2) = (0.8, 0.2)$, we have $^C D_\gamma^{\alpha(\cdot, \cdot), \beta(\cdot, \cdot)} x(0.4)$ ≈ 0.7144:

```
a = 0; b = 1; n = 1;
alpha = @(t,tau) (t.^2 + tau.^2)/.4;
beta = @(t,tau) (t + tau)/3;
x = chebfun(@(t) t, [0 1]);
gamma1 = 0.8;
gamma2 = 0.2;
CC = combinedCaputo(x,alpha,beta,gamma1,gamma2,a,b,n);
CC(0.4)
ans = 0.7144
```

For other values of $^C D_{\gamma^n}^{\alpha(\cdot, \cdot), \beta(\cdot, \cdot)} x(t)$, for different values of $t \in (0, 1)$ and $\gamma = (\gamma_1, \gamma_2)$, see Fig. A.4 and Table A.2.

References

1. Driscoll TA, Hale N, Trefethen LN (2014) Chebfun Guide. Pafnuty Publications, Oxford
2. Linge S, Langtangen HP (2016) Programming for computations-MATLAB/Octave. A gentle introduction to numerical simulations with MATLAB/Octave, Springer, Cham
3. Trefethen LN (2013) Approximation theory and approximation practice. Society for Industrial and Applied Mathematics

Index

Printed in the United States
By Bookmasters